SERIES OF CONTEMPORARY ARCHITECTS STUDIO REPORT IN THE UNITED STATES

Jennifer Siegal

Contents

Jennifer Siegal

DESIGN
SIX WORKSHOPS
FOR 14-17 YEAR OLDS
JUNE 23-27 2003
GEARING
MOVING
WEARING
MORPHING
EATING
DWELLING
CAMP
MOBILE
THE ART OF
PORTABLE
ARCHITECTURE
PANEL #1
ROLLERBLADE

Contents

Jennifer Siegal, The Best and Brightest

ESQUIRE December 2003

Architecture that doesn't move–physically pick up and move–is just so second millennium. We're living in a mobile world, and if not just individuals but entire communities can't keep up when they need to, then what good are they?

建筑不会动，无论是升高还是移动，一千年也不会有变化。今天，我们生活在一个移动的世界里，作为个人和整个社会都需要跟上潮流的步伐。

JENNIFER SIEGAL is fed up with architecture that can't fit on the back of a flatbed truck. We live in a world of cell phones, PDAs, in-car e-mail, and Triscuit-thin Internet-equipped laptops. Our existences are increasingly mobile, as our odometers attest, but architecture has remained stubbornly anchored to the ground. If Siegal has anything to do with it, that's about to change. Jennifer Siegal and her Los Angeles-based firm, office of Mobile Design, have invented the trailer home of the future: cheap, high-tech, eco-friendly, and–here's the shocker–pretty to look at. Erase from your mind the trailers you've seen behind pantsless drunken guys on Cops. Siegal's trailers–called Portable Houses–are Pleasantly light and airy; the walls are made of special translucent plastic with air pockets that act as insulation. The other building materials sound like something from the menu of a macrobiotic restaurant: compressed sunflower-seed hulls and Plyboo, a good–for–the–environment bamboo product. And Portable Houses are surprisingly cheap: Each is pieced together in a factory from prefab parts, which means a 720-square-foot model runs about $80,000.

珍尼弗·西格尔厌烦了装不上拖车的建筑。我们生活在一个充满手机、商务通、车载电子邮件和能上网的超薄笔记本电脑的世界。人类的生存变得越来越充满动感，我们车中的里程表能够证明这一点。然而，我们的建筑依然顽固不化地钉在地面上。西格尔要对此做出改变。她和她在洛杉矶的移动设计事务所（Mobile Design）已经成功发明了未来的房车——便宜、高科技、环保，而且令人惊奇的是外观非常漂亮。那种住着衣衫不整的醉鬼的房车也许将会从你的记忆中消失。西格尔房车被称作是便携式房屋，非常轻便通风；墙壁由配有隔热气囊的特殊透明塑料制成。其他的材料听上去就像来自一个益寿延年的菜谱，压缩向日葵籽壳、环保竹产品。便携式房屋的价格惊人的低廉，一个720平方英尺的房子只需80 000美元。

Siegal–who grew up in New Hampshire, spent her teen years in Israel, and worked a mobile hot-dog stand to help pay for architecture school-has a grand scheme for the future: cities in which any empty lot is fair game for a mobile unit to park, plug in, juice up, and pull out. It's a world for the growing group she calls the New Nomads.

西格尔在新罕布什尔州出生，在以色列度过其少女时代。她曾经在一个活动热狗摊工作以支付在建筑学院的学费，这为她将来的设计打下了基础。城市中任一空间都给予移动房屋停驻、安装、活跃和离去的机会。她称这是一个新流浪者的世界。

Her first mobile-home village will soon become a reality–an artists' community called Eco-Ville. A handful of what will be forty Portable Houses will make the trek forom a factory in Rancho Cucamonga, California, to the 2.5-acre plot in downtown L.A. Siegal designed a special prototype for Eco-Ville, a set of two modular units stacked to allow a roof garden on one side and a shaded garage on the other. These homes will be fairly permanent once they take up their places in Eco-Ville, a seeming

她的第一个房车村落将诞生，一个称作生态村的艺术家社区。40辆房车将从加州的一个工厂中生产出来，运往洛杉矶市中心的一个占地2.5英亩的地区。西格尔为生态村设计了一个特别的原型：两个挨在一起的单元，一边有顶楼花园，另一边有遮蔽式车库。这些房屋将永久

Rebecca Dorr

珍尼弗·西格尔
最出色和最聪明的建筑师

contradiction that doesn't have Siegal all that worried: "It's about living lighter on the land with smarter architecture."

The Portable House is just one in a line of Siegal's mobile buildings. The Mobile Eco Lab, an early project, turned a retired cargo trailer into an environmental classroom on wheels. For the Haagen-Daz Pleasure Mobile, Siegal designed a portable ice-cream-bar-cum-mobile-theater in which you can order sorbet cocktails and screen films. In the future, she may escape the unmoving nature of land altogether: Her newest designs are for technologically advanced homes–for the ocean.

性驻扎在生态村，这是一个看上去有些矛盾的形式。但是西格尔并不担心，她说："这是更聪明的建筑，它是地球上的生命之光。"

便携式房屋只是西格尔设计的移动建筑之一。在较早的移动生态实验室项目中，西格尔把一个废弃的货箱转变成环保的轮子上的教室。西格尔还设计了哈根达斯移动售货服务站，那是一个移动的冰激凌吧兼电影院，你可以在里面点果汁、冰糕和鸡尾酒，也可以欣赏最新的电影。在将来，西格尔也许会逃离不动的大陆。她最新的高科技设计将在海洋中实现。

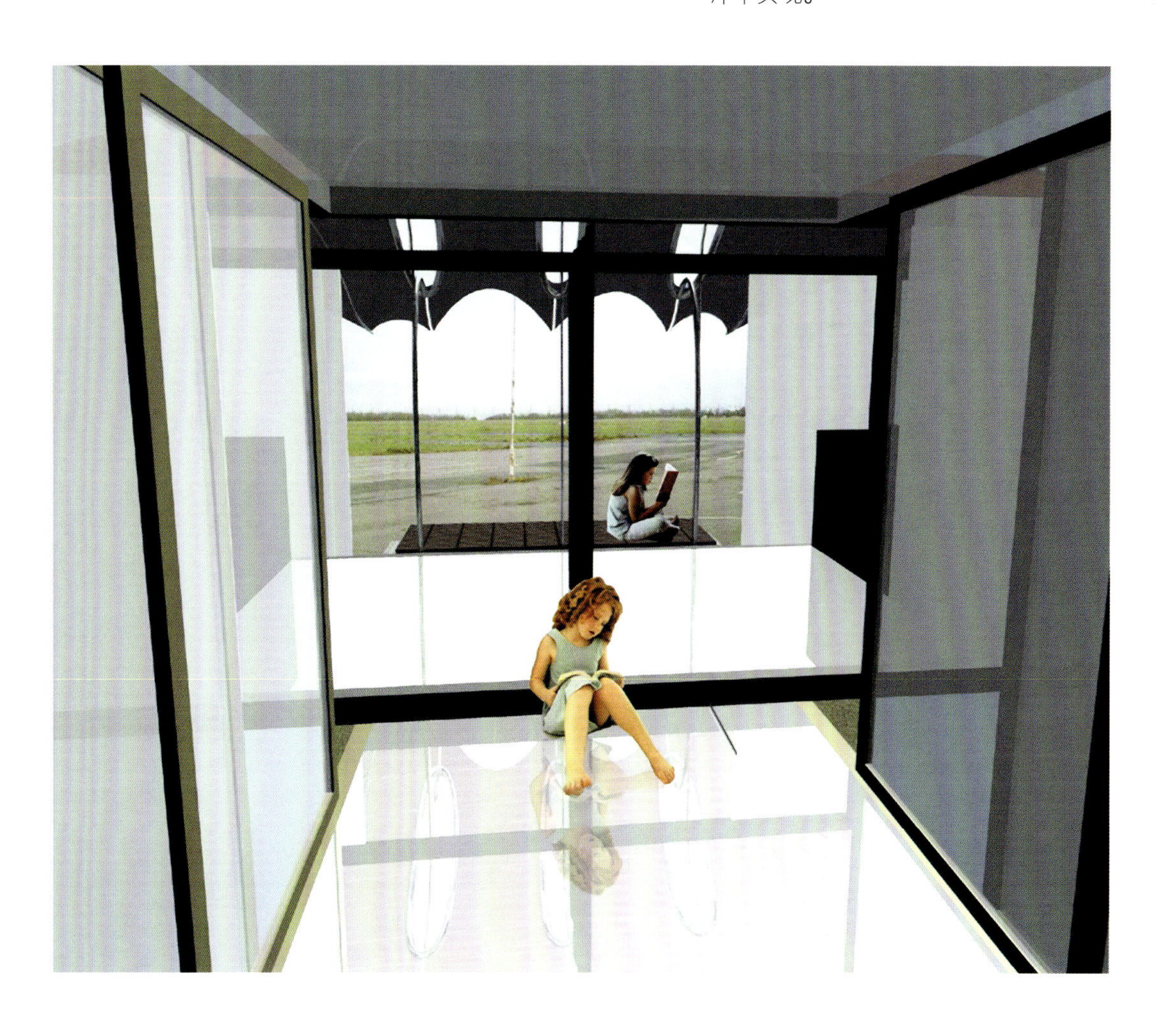

Right: rendering of mobile event city
移动城市效果图

The New Mobility

"HOME IS WHERE YOU PAER IT." So goes the bumper-sticker philosophy stuck on recreational vehicles motoring down America's highways. It's not entirely trite: Some 2.8 million Americans are estimated to be fulltime Rvers, a vast mobile population that, until recently, the U.S Census didn't quite know how to classify (much less find), simply labeling its members the "affluent homeless."

The bulk of these nomads would not profess to be part of some avant-garde design movement, and the houses-on-wheels they pilot don't capture the imagination of most architects. These barriers may be beginning to fall. There's a new energy in the field of mobile ity and prefabrication, driven by new materials and technologies, the changing social conditions imposed by the 24/7, "just-in-time" global economy and an expanding interest among various demographic groups in a mobile lifestyle. As Jennifer Siegal, head of Siegal Office of Mobile Design(OMD)in Venice, Calif, and one of the most prominent of the new mobility proponents, writes in her book, mobile: The Art of portable Architecture (Princeton Architectural Press, 2002), "Architecture today rolls, flows, inflates, breathes, expands, multiplies and contracts, finally hoisting itself up, as Ö predicted in the early 1960s, to go in search of its next user."

This movement isn't just a string of verbs on paper. Active members include FIL Design Engineering Studio, which specializes in lightweight and deployable constructions such as the Carlos Moseley Music Pavilion, a traveling performance venue designed for a number of New York cultural institutions. Concocting a fabric for the pavilion, whose shape is based on acoustic and sheltering needs, the firm raises the Corbusian question: "When is a structure a machine and when is it a building." Then there's the Mark Fisher Studio, a group responsible for the Rolling Stone's Steel Wheels set–a vast, post-industrial apparatus that represents the largest-ever traveling stage. The firm Festo has created what it calls "the first building in the world to be constructed with a cubic interior"(supporting structure built with air-inflated chambers). The humble steel or aluminum shipping container, the ubiquitous conveyance of globalization, is now appearing in so many projects–from LOT/EK to OMD to Wes Jones Partners–that

Tom Vanderbilt

“你停靠的地方就是你的家。”这贴在穿梭于美国高速公路上的休闲车车尾标签上的口号是对移动人群的最好诠释。估计有280万美国人是这样生活的。这是一个巨大的移动人群。直到现在，美国的调查机构也不知道该如何给这些人群分类，只是简单地称其为“富足的流浪汉”。

大量这样的流浪者未曾能吸引前卫的设计运动，轮子上的房屋也不曾吸引建筑师的目光。而现在，这些障碍正在被打破，移动建筑领域出现了一股新的力量。对新材料和新技术、变幻莫测的社会和经济环境，以及对拥有不同生活方式人群的兴趣驱使建筑师们不无兴奋地重新表达移动建筑的概念。加利福尼亚州威尼斯动感设计事务所(OMD)的总设计师珍尼弗·西格尔，一位著名的新移动建筑的拥护者，在其所著的《移动：便携式建筑的艺术》（普林斯顿建筑出版社2002年出版）一书中这样说：“今天的建筑能够滚动、飘扬、膨胀、呼吸、扩展、复合和缩减，最后再升起来，就像20世纪60年代人们预测的那样，去寻找它的下一个使用者。”

这种运动并不只是一纸空谈。积极从事此类活动的还包括FTL设计工程工作室。他们专门从事轻型的可展开建筑的设计工作，曾成功地为众多的纽约文化机构设计了一个称作卡罗斯莫斯利音乐亭的旅行表演舞台。亭子的形状根据听觉和遮蔽的需要来设计，这家工作室提出了这样一个“柯布西耶式（Corbusian）”的问题：“结构什么时候是一架机器，什么时候是一座建筑？”后来，马克·费舍尔（Mark Fisher）工作室设计了滚石钢轮系列，一个巨大的后工业设备：最大的旅行舞

最新的移动

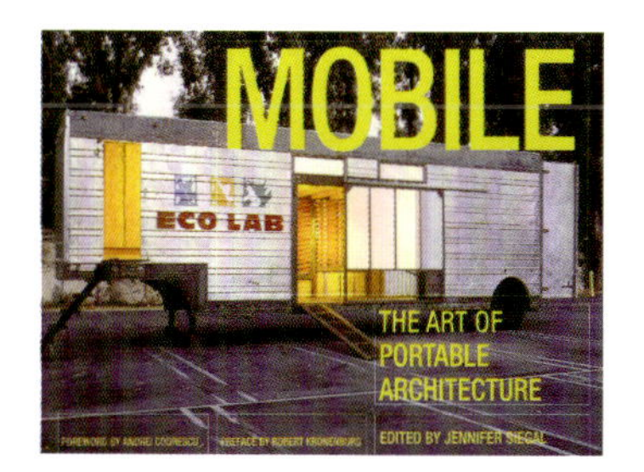

it doesn't seem gratuitous to wonder, as pre fab(Gibbs Smith Publisher, 2002) authors Allison Arieff and Bryan Burkhart do. Whether it might become the 21st-century brick, which is what Paul Rudolph suggested the mobile home would be for the previous century.

The idea of mobile architecture and its corollary, prefab architecture, has long haunted the dreams and sketchbooks of the most visionary architects-from Buckminster Fuller's 1929 Dymaxion house, whose components could be reduced to a tube (eight of which could fit on a standard railway car). To Frank Lloyd Wright's inflatable rubber Airhouse, built for U.S. Rubber at the 1959 International Home Exposition. Due to technological flaws, production costs or lack of popular acceptance, these schemes rarely made headway. Most mobile architecture has emerged on the fly out of economic or strategic necessity (like the Quonset hut developed at the eve of World War II by the George A. Fuller Construction Co.), or by nomadic societies (the igloo, for example, can be fashioned in one hour by a skilled Inuit using the packed snow from a single snowfall.)

Arieff and Burkhart point out that "the majority of housing built in the United States and built in the United States and abroad is, to some degree, prefabricated." So are most of the commercial structures–the fast-food chains, big-box stores, motels–that define the new American settlements. They note that most of these building types are architecturally uninspired–simply ersatz versions of traditional "stick-built" structures–and nonspecific, breeding an instant homogeneity and monotony, the 1960s dream of avant-garde architectural collective Archigram's "Plug-in Cities come back as placeless night mares. There's no theory attached to these instant landscapes; rather, they're simply a physical expression of the most brutalis-tic economic imperatives.

"Mobile homes" (a misnomer because more than 90 percent of such structures are mobile only once–from factory to housing location) are becoming a landscape fixture particularly in the fastest-growing regions of the country. According to the 2002 State of the Nation's Housing report from Housing Studies "Manufactured housing was responsible for 35 percent of the

台代表。费斯通（Festo）创建了被称作是“世界第一个在立方体内部建造的建筑”（在一个充气室里建造的支架结构）。在全球化运输中很普及的钢制铝制集装箱，现在出现在很多项目中，从“LOT/EK”到“动感设计事务所”再到“威斯琼斯合伙人”。《便携式构造》一书的作者阿里森·阿瑞夫（Allison Arieff）和布莱恩·伯克哈特（Bryan Burkhart）都深信不疑地认为可移动建筑是21世纪的主导。保罗·鲁道夫曾认为这种建筑在上个世纪就已经绽放无限光彩。

移动建筑以及它的分支——组装式房屋的概念一直缠绕在建筑师的梦想中和写生簿里。在1929年巴克敏斯特·福勒（Buckminster Fuller）的节能住宅项目里，建筑部件被减少到一个小包厢（8个这样的包厢能够安装成一个标准铁路客车）。1959年的国际房屋展中，弗兰克·劳埃德·赖特（Frank Lloyd Wright）为美国橡胶公司建造了一个橡胶充气屋。但因为技术的缺陷，制造成本或者公众的接受程度问题，这些几乎都进展不顺利。大多数移动建筑的出现都缘于经济和战略的需要，比如二战前期乔治·A·富勒建筑公司开发的匡西特活动房屋(一种用预制构件搭成的长拱形活动房屋)。还有游牧社会建造的类似房屋，比如，因纽特人能够在一个小时的下雪时间内用积雪建起一座圆顶建筑。

阿里森·阿瑞夫和布莱恩·伯克哈特指出：“美国和国外大多数建筑在一定程度上是预制安装的。”很多商业房屋也是如此，快餐连锁店、仓储式超市和汽车旅馆定义着新的美国建筑风格。他们表示，大多数这类建筑是缺乏创见的，它们只是传统拼合组装建筑的翻版。这

growth in homeownership in non-metropolitan areas and 23 percent of the gains among very low-income households." Mostimpressively, the report noted, "Manufactured housing's share of growth in the South was 30 percent overall and fully 63 percent in rural areas."

The manufactured house, now representing two out of every IO new home purchases overall, hasn't entirely managed to shed the perception that its popularity is limited to two demographic categories–the newly wed or the nearly dead. After all, many countries have a long-standing cultural bias against nomadism–property confers legitimacy, social standing, even voting rights. As the catalog for the recent Vitra Design Museum exhibition Living in Motion puts it, "The distinction that our culture has made since biblical times between nomads and settlers has resulted in our exclusion of huts tents or igloos from our conception of architecture, just as we don't regard baskets, hammocks or pillows to be furniture. "The bulk of manufactured housing, moreover, still feels dictated more by the exigencies of the assembly line than by appreciations of space and movement. And yet, as the writer J.B. Jackson, in his essay "The Mobile Home on the Range." Once suggested, "It almost seems as if these shortcomings, which the critics never tire of mentioning–the lack of individuality, the functional incompleteness, the dependence on outside services and amenities, and even the lack of traditional architectural qualities as firmness, commodity and delight–are what make the trailer useful and attractive to many of its occupants."

Why, despite the social bias, is there a new turn toward mobility? Notably, the phrase "mobile architecture" most commonly refers to the net-working systems of mobile phones and PDAs. The semantic coincidence. "While information has been developing at a rapid rate–cells, portables, laptops, imagery–the space in which these experiences occur was not being looked at," Siegal says. "The slow of information was leaving the flow of the body experience behind." The proliferation of Wi-Fi is bringing a new ubiquity of electronic communications coverage to spaces–from entire cities to the single home–before architects and planners have had a chance to consider how this flurry of chat might reorder environments. As Siegal notes of a recent trip to

种千篇一律、没有特点的建筑非常令人失望，感觉就像是20世纪60年代的前卫建筑阿基格拉姆的“插入式城市”的噩梦又回来一样。这些即时建筑不带有理论基础，只是大多数经济疯狂发展的物理表达。

“移动的家”（一个不当的用词，因为9.0%的此类建筑总共只会从工厂到目的地移动一次）正在成为一种景点装置，特别是在国家发展最快的地区。根据哈佛大学房屋研究联合中心在2002年的全国房屋报告：“预制式住宅促进了非城市地区房屋拥有量35%的增长和低收入家庭23%的收入增长。”令人印象最深的是，报告指出“南部预制式住宅的增长达到30%，其中农村地区占了63%。”

10栋新房子中有2栋是预制式住宅，但是这不能完全掩盖这个事实——它主要在新婚或者老年人群中比较受欢迎。总之，很多国家对流浪主义有很大的偏见，他们认为不动产赋予了人的合法性、社会地位甚至选举权。最近，威特拉设计博物馆的展览“移动中的生活”这样表示：“从游牧时期到安居时期，我们的文化中显现的不同使人们将木屋、帐篷或者圆顶房屋从自己的概念中排除掉了，就像我们不再把篮子、吊床或者枕头看作是家具一样。”规模巨大的预制式住宅依然受到生产线的主导，而不是出于对空间和运动的欣赏。作家J·B·杰克逊在他的文章《并列的移动房屋》中说：“下列缺点总是受到批评：缺少个性、功能的不完善、依靠外界的服务和娱乐设施，甚至缺少传统建筑的坚固、舒适和方便”。尽管如此，移动房屋对很多居住者来说是有用而且吸引人的。

Cambodia and Laos: "Guess what? Internet cafes in the jungles and in cities with no infrastructure."

Mobility, it might be said, is here to stay, While for some it might represent the ultimate freedom of choice, it's the only choice of those who have lost their freedom. The English firm LDA, winner of a competition sponsored by Architecture for Humanity and the United Nations High Commission for Refugees, created a mobile prefab shelter (motivated by the war in Bosnia) that uses local materials–including rubble. "Our shelter was designed to be constructed with bare hands for situations where there's unskilled labor and no tools," LDA partner Mike Lawless says. "It's secure, wind-and watertight, which would allow people to live reasonably while they reconstruct their house.. The idea is that the system is a product, one from which they can solve the problems of their particular location."

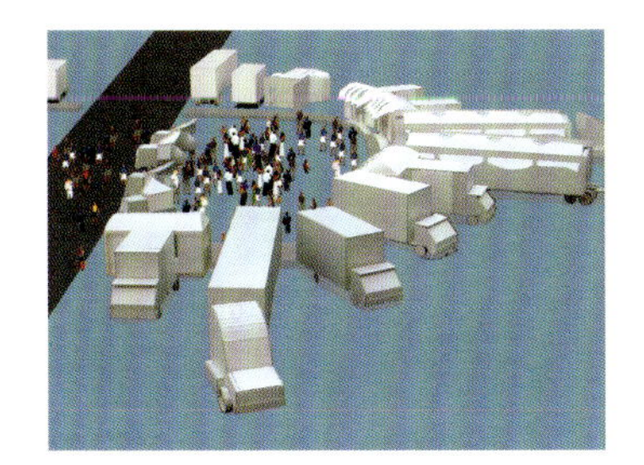

Architecture for Humanity recently oversaw another mobile-architecture competition, a mobile AIDS clinic designed for use in sub-Saharan Africa. Siegal, who served as a juror, says that the designers of many submissions seemed to cleave to ideas of static architecture simply transported, rather than creating structures whose very form derived from mobility. "It was hare to convince the more traditional architects that a purely mobile structure could be built at a reasonable cost, perform all the duties that it needed to and be an interesting place to work out of and experience." She says. "The project I liked the best–Bubblebug–was terrific, but not taken seriously by many of the jury members, as they couldn't understand the technology-how it was cooled, heated inflated, etc."

Extreme mobility tends to characterize either the very poor or the very rich–the transnational migrants (sometimes, ironically, hiding in shipping containers) seeking better wages, or the permanent leisure class living onboard The World, a cruise ship with permanent living units, which constantly sails the globe (recalling Fuller's speculative floating "Triton City"). Others are investigating mobility as a way of life. Some see the annual countercultural Burning Man Festival in Nevada as a kind of Archigram-like "instant city." But given its temporary nature like-minded group

除了社会偏见，对于移动的概念有没有一种新的趋势呢？很明显，“移动建筑”最常指手机和商务通的网络工作系统。语义上的类似其实不仅是个巧合。西格尔这样说：“随着信息迅速地发展，手机、商务通、笔记本电脑等工具层出不穷，然而出现这些事物的空间却不被人重视。信息流正在把身体体验的潮流落在后面了。”在建筑师和规划者有考虑环境问题的机会之前，Wi-Fi无线技术的发展为生活空间——无论是整个城市还是单个家庭都带来了普及的新的电子化交流。西格尔谈到了最近到柬埔寨和老挝的旅行经历：“你猜怎么着？网吧都在丛林里，城市里根本都没有基本设施。”

移动，也许有人会说就呆在这里。对一些人来说它代表着最高的自由，它是那些失去自由的人的惟一选择。人道主义建筑组织和联合国难民救济总署资助的人道主义建筑设计竞赛冠军——英国公司LDA创建了一个移动预制式避难所（因波斯尼亚战争），他们使用了当地的材料，包括橡胶。LDA的合伙人迈克·劳里斯（Mike Lawless）说：“避难所的设计考虑到安装时需要徒手而没有任何工具的情况。这很安全，不漏风不漏雨。当人们重建他们的房子时，能够住得合理舒适。该系统是通过这个产品，使人们能够解决特殊环境的问题。”

人道主义建筑组织最近参与了另一个移动建筑竞赛，是为撒哈拉沙漠的非洲地区建造的移动艾滋病诊所。作为评委的西格尔表示，很多设计者似乎只会将静止的建筑简单地做成可以运输的方式，而不是创造形式源于运动本身的建筑。她说：“传统的建筑师总是不相信：一

of residents, inhospitable state-owned surroundings (the citizens must live on imported supplies), Burning Man is a mirage of a city–a place of fleeting reinvention that strangely parallels its glittering pro-consumerist neighbor to the south–that would collapse on its own internal contradictions.

A more realistic evocation of what a mobile community might look like is found in California, in "Slab City," a collection of trailers and other mobile structures parked on the concrete slabs of a former Army base located in the desert. An amorphous, shifting collection of transients and "residents"–a town but not a town–Slab City is off-the-gird urbanism existing on state-owned land, a place where people have gathered to get lost. Similarly, artist and designer Andrea Zittel has personified of the West with her "A–Z Homestead Unit." Smaller than 120 square feet and built on five-acre plots near Joshua Tree, Calif. (originally given free by the government to those who would "improve" them), the temporary structures are considered outside of zoning laws.

It has been more than a century since the American frontier was declared closed, thus putting a putative end to our dreams of endless expansion and unfettered mobility. At the same time, many of us continue to dwell in a kind of accidental mobility–we build new fixed constellations of ex-urban sprawl that we'll inhabit only temporarily, while our cars, presumed harbingers of mobility, begin to resemble houses, with their in creased interior spaces (including over sized cup holders designed to hold food containers), proliferating entertainment and communication options, and increased "dwell time." Mobility today isn't as simple as roaming unconquered frontier or hitting the open road for exotic new destinations. Mobility today is about responding to the pulse of global capitalism, about inhabiting the "space of flows," those interstices of money, commerce and information–or even about slipping between the cracks, but we continue to adhere to illusions of permanence. The architects of the "New Mobilism" are working to give a new shape to this movement, creating structures that guarantee that no matter where you are, you're always home.

(Written by Tom Vanderbilt, the contributing editor of I.D.)

个真正的移动建筑的造价可以非常合理、低廉，能够拥有所需的所有功能并且是一个有趣的工作和体验的地方。我最喜欢的作品是Bubblebug。它真是太棒了。但是其他很多评委并不重视它。因为他们不能理解其中的技术：如何制冷、供热和充气等等。

极致的移动者既有穷人也有富人，寻找更好薪水的跨国移民（有时很讽刺地藏在集装箱里），以及生活优裕的上层人士经常出海旅游（想想福勒设计的华丽、浮游的“海神城”）。另外的人把迁移作为一种生活的方式。一些人把内华达州野蛮的焚人节看作一种具有Archigram(20世纪60年代在英国崛起的“披头六”建筑团体)派特点的“即时城市”。但是考虑到它的暂时性质、居民的相似思维、周围不友好的环境（居民以进口物资为生），焚人是这个城市的幻境，一个短暂的彻底改造一切的地方，它奇怪地使其与南部强大的邻居取得了一种平等，同时也解决内部的纷争。

一个更现实的体现移动社区特色的实例在加州被发现。这是个“板城”。一系列房车和其他移动装置被停放在沙漠中某个过去驻军基地的水泥板上。一个无定形的、游离的瞬间和“居民”：一个小镇但又不是真正的小镇。“板城”脱离了存在于国有土地上的城市主义。在这片土地上，聚集的人们迷失了方向。与其类似，艺术家兼设计师安德里亚·基特尔（Andrea Zittel）在她的“A-Z家园”中将西部游牧主义和自给自足的生活方式人格化。这座建筑占地小于120平方英尺，建在临近加州约书亚树附近的5英亩的土地上（该地区的人被政府给予自由，因为他们能够

“改善”自己）。临时的建筑被置于地区法律之外。

自美国边境宣布关闭以来已有一个世纪。我们进行无尽扩张和不羁迁移的梦想也就终止了。同时，我们中的很多人继续进行意外的移动，建立新的向城市外延伸的格局。我们只临时性地居住，我们的汽车预示着迁移，我们开始组装房屋并充分利用它们的内部空间（包括设计用来支撑食品容器的特大杯座），我们创造更多的娱乐和交流方式并增加“停留时间”。如今的移动并不是像在边境漫游或者沿着马路寻找新目的地那么简单了，而是一种对全球资本主义发展脉搏的反应。它与“流动空间”中的生活，那些金钱、商业和信息的空隙，甚至是空隙之间的移动有关。我们现在都是流浪者，但是我们也继续坚持对永恒的幻想。“新移动主义”的建筑师正在努力塑造这个运动的新外形，创造一种能够保证无论你在何处，总是在家的建筑。

（本文由美国I.D.杂志主编汤姆·范德比尔特撰写）

Left: the view of Jennifer Siegal's exhibition at Harvard University

珍尼弗·西格尔在哈佛大学的作品展览现场

Contents

omd
omd

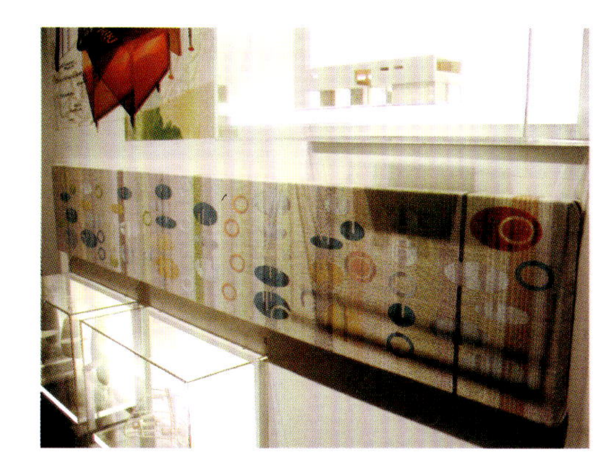

Left: exhibition view at Cooper Hewitt Design Musem

珍尼弗・西格尔在库珀・休依特国家设计博物馆的作品展览现场

Storehouse

Completion Date: National Design Triennial 2003
Client: Cooper-Hewitt, National Design Museum, Smithsonian Institution
Principal: Jennifer Siegal
Senior Designer: Kelly Bair
Fabrication: Penwall Industries
Smart Skin: International Fashion Machines,
Joanna Berzowska, CCO
Maggie Orth, CEO

Constructed as a mass-customized modular unit, and built from titanium with scrim/fabric clad wings, Storehouse displays architectural models and drawings in an intricate system of hard and soft materials. The wings armature is randomly punctured with shadowbox shelves, creating depth so that each project displayed is uniquely perceived. The base acts both as anchor and

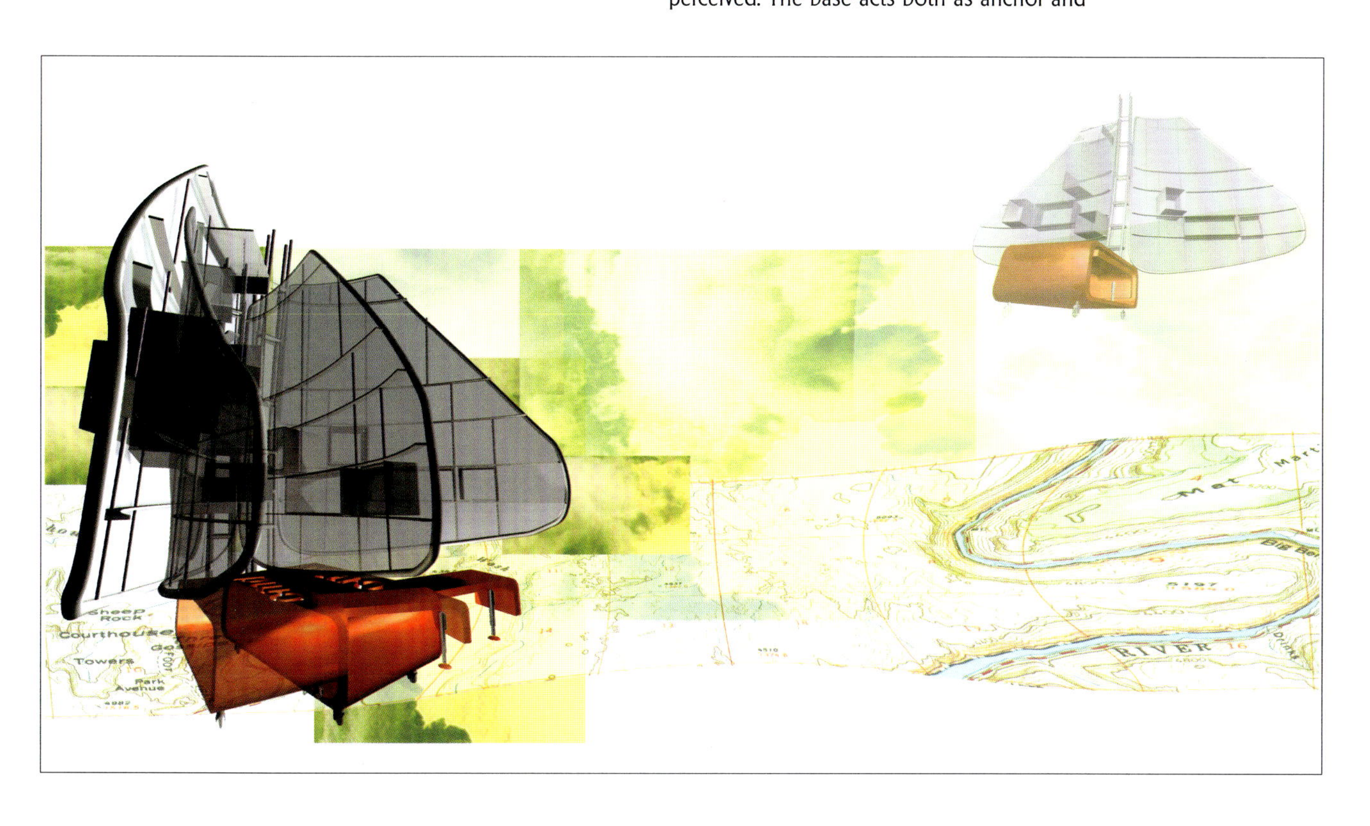

Opposite and left: Mobile book kiosk concept rendering for installation

移动书刊亭装置效果图

Right: exhibition view at Cooper Hewitt Design Musem

珍尼弗·西格尔在库珀·休依特国家设计博物馆的作品展览现场

seating bench, unfolding, rolling and adapting to provide solace for museum visitors.

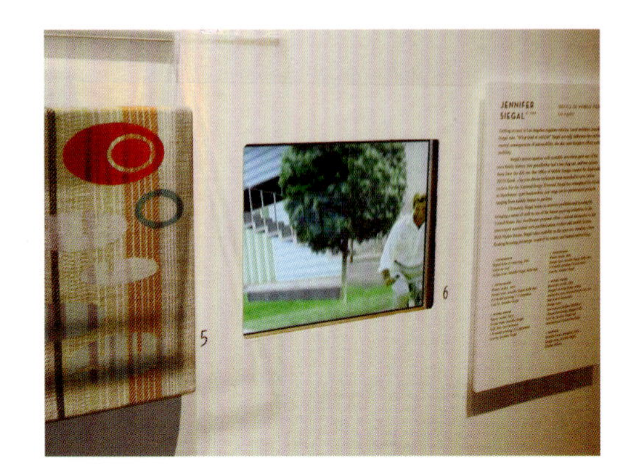

Below: Mobile book kiosk concept rendering for installation

移动书刊亭装置效果图

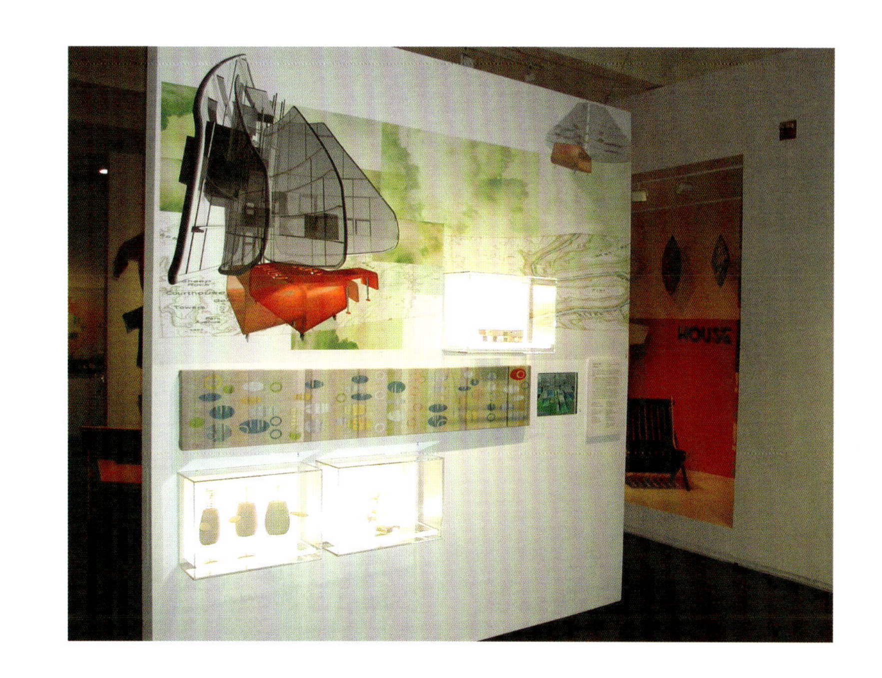

店铺项目

国家设计博物馆，华盛顿，2003

这是一个为国家设计博物馆特制的合成材料展示设施，由钛合金构成，并带有覆盖着棉麻织物的两翼。它通过一个复杂的软硬材料系统来展示建筑模型和图画。两翼的防护罩会随机展开，让每一个被展示的项目都产生独特的感知。底部既有固定作用，又是休息的长椅，它伸展、摇摆并亲和地为博物馆的来访者提供慰藉。

Contents

移动生态实验室

洛杉矶好莱坞，加利福尼亚州

移动生态实验室是与“好莱坞美化团队”——一个旨在保护好莱坞社区美丽和完整的绿化组织合作建造的。项目通过使用电脑动画、传统建筑制图和大型建模技术的方式进行语言和视觉的交流。整个工作通过明确定义的材料来体现（一个捐赠的货运拖车）。这个8英尺×35英尺的房车现在已经巡回穿越洛杉矶，向这一地区的学校的学生们宣传挽救和保护地球的重要性。

作为一个移动教室，移动生态实验室展示了一系列有关生态环境的焦点问题。一个关于“树的生命”的多媒体课程在这个能够扩展的移动生态实验室中创造了一个内外交织的探索通道。同时这也是一个可由当地艺术家引导孩子们创作的，并以此替代校园涂鸦的艺术工作室。老师们可以利用这个像舞台一样的讲台来与孩子们探讨植树和保护环境的重要性。

这个标志性的移动建筑就像一个马戏团一样穿梭于校园中，在那里，升高的引道被折叠好并滑离拖车。孩子们立刻就能认出这个充满交流、发现和乐趣的地方。

Mobile Eco Lab

Completion Date: June 1998
Owner: Hollywood Beautification Team, Sharyn Romano, Director
Principal: Jennifer Siegal
Woodbury University Project Team: Ausencio Ariza, Larry Cheung,Thomas Cohen, Tinifuloa Grey, Chayanon Jomvinya, Jody Segraves
Assistants: Alex Arias, Guillermo Delgadillo, Han Hoang, Maurice Ghattas, Thao Nguyen, Jose Olmos, James Popp, Phung Thong, Juan Uehara

The mobile Eco Lab was built in collaboration with the Hollywood Beautification Team, a grassroots group founded with the mission to restore beauty and integrity to the Hollywood community. Verbal and visual exchanges took place using computer animated drawings, traditional architectural drafting, and large scale modeling techniques. Full-scale work was performed with a defined material palette (specifically that of a donated cargo trailer and cast-offs from film sets). The 8 x 35 foot trailer now travels throughout Los Angeles County to inform K-12 school-aged children about the importance of saving and protecting our planet.

As a working mobile classroom, the Eco Lab provides a base for a range of exhibitions all of which focus on ecology. A multimedia program explaining the "life of a tree" creates a path for discovery that weaves in and out of the

expandable Eco Lab. A working art studio, local artists collaborate with the children to create facade-sized murals replacing graffiti at inner-city schools. School teachers use stage-like platforms to discuss each child's role in the importance of planting trees and maintaining a sustainable environment.Like a circus tent, this mobile icon arrives at the schoolyard where elevated walkways fold down and slide out of the trailer's body. It is immediately recognizable as a place for interaction, discovery and fun.

Below: construction view of Mobile Ecolab and dusk view of completed Ecolab

移动生态实验室施工现场和黄昏中的生态实验室

Opposite and below: Ecolab interior view and Ecolab modeling construction sequence photo

生态实验室室内、模型和施工现场

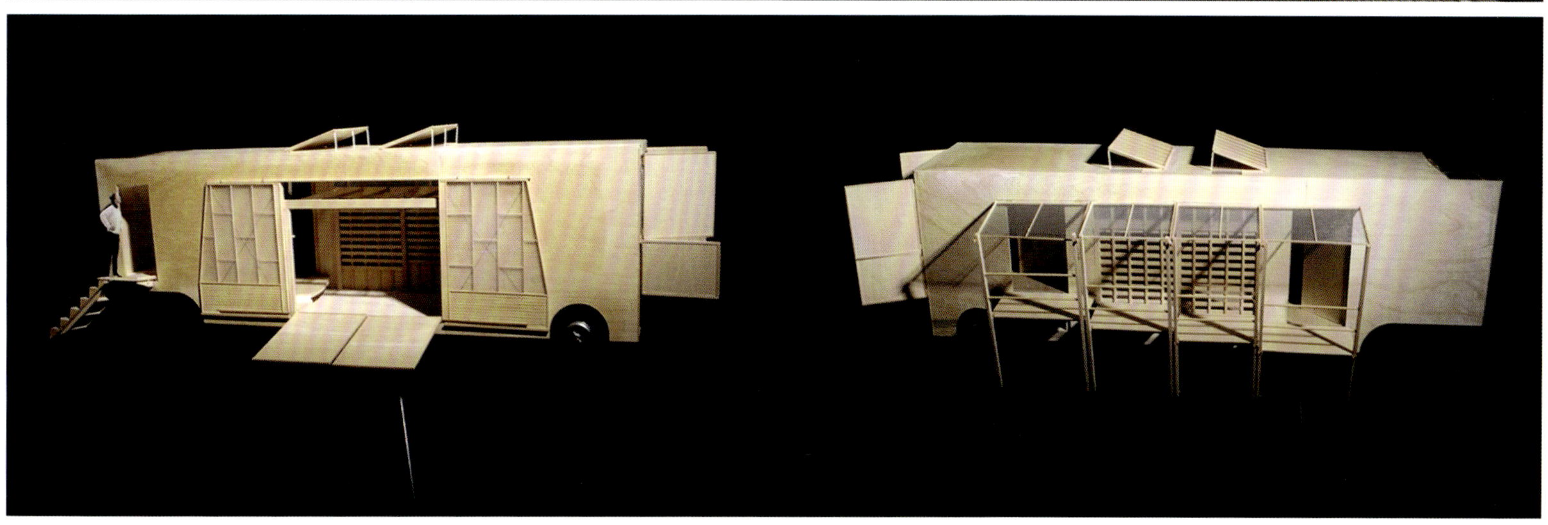

Contents

Eco-Ville

Completion Date: pending
Clients: Tom Ellison
Principal: Jennifer Siegal
Design Team: Kelly Bair, Sara Schuster, Andrew Todd

Transforming a 2 1/2-acre lot in downtown Los Angeles is the focus of the Eco-Ville Development project. The program involves the development of Artist-in-Residence live/work space. Approximately 40 Portable House units are to be deployed at Main Street in Los Angeles, California.

The project is to develop and construct a sustainable Artist-in-Residence live/work community. The final objective is to construct and deploy multiple versions of the Portable House (a mass-customized residential building unit), with an emphasis on native Californian drought resistant plant materials, common gardens, and the use of sustainable building materials. The Eco-Ville Development is comprised of a series of attached and semi-attached buildings in multiple stacked configurations. The bottom unit provides a flexible work space, and the attached upper unit offers an open well-lit living space with access to a private roof garden. In an effort to provide affordable artists' residences, the development demonstrates that individual modern design solutions are possible with mass-customization.

生态村

洛杉矶，加利福尼亚州

洛杉矶市中心的一个占地2.5英亩的艺术家生活/工作地的空间改造项目是生态村开发项目的工作焦点。大约40座房屋将被安装在加州洛杉矶的中心大街。这个项目要开发建造一个具有可持续性开发的驻村艺术家生活/工作社区。最终的目的是建造和安装不同类型的移动房屋（一种大型的移动合成建筑），着重于加州原产的抗干旱植物材料、普通花园，以及对可持续性建筑材料的使用。生态村的开发包括一系列不同类型的附加和半附加建筑。底层提供一个灵活的工作空间，附加的高层部分则提供了一个开放的、阳光充足的生活空间，还可以通向一个私人屋顶花园。该项目旨在为艺术家们提供经济的居住地。这个开发项目示范了个性化的现代建筑构造方法，也能应用于大规模建筑定制。

Opposite and next two pages: rendering of Eco Ville, the inexpensive, high-tech, eco-friendly modular homes are intended for a growing group, Siegal calls the new nomdas and are designed for frequent, easy movement anywhere on earth

生态村效果图，西格尔认为这些由廉价材料、高科技和生态标准件组成的住宅是新游牧主义的理想模型，并可在地球上任何地方自由移动

Left: inspirational image

灵感来源

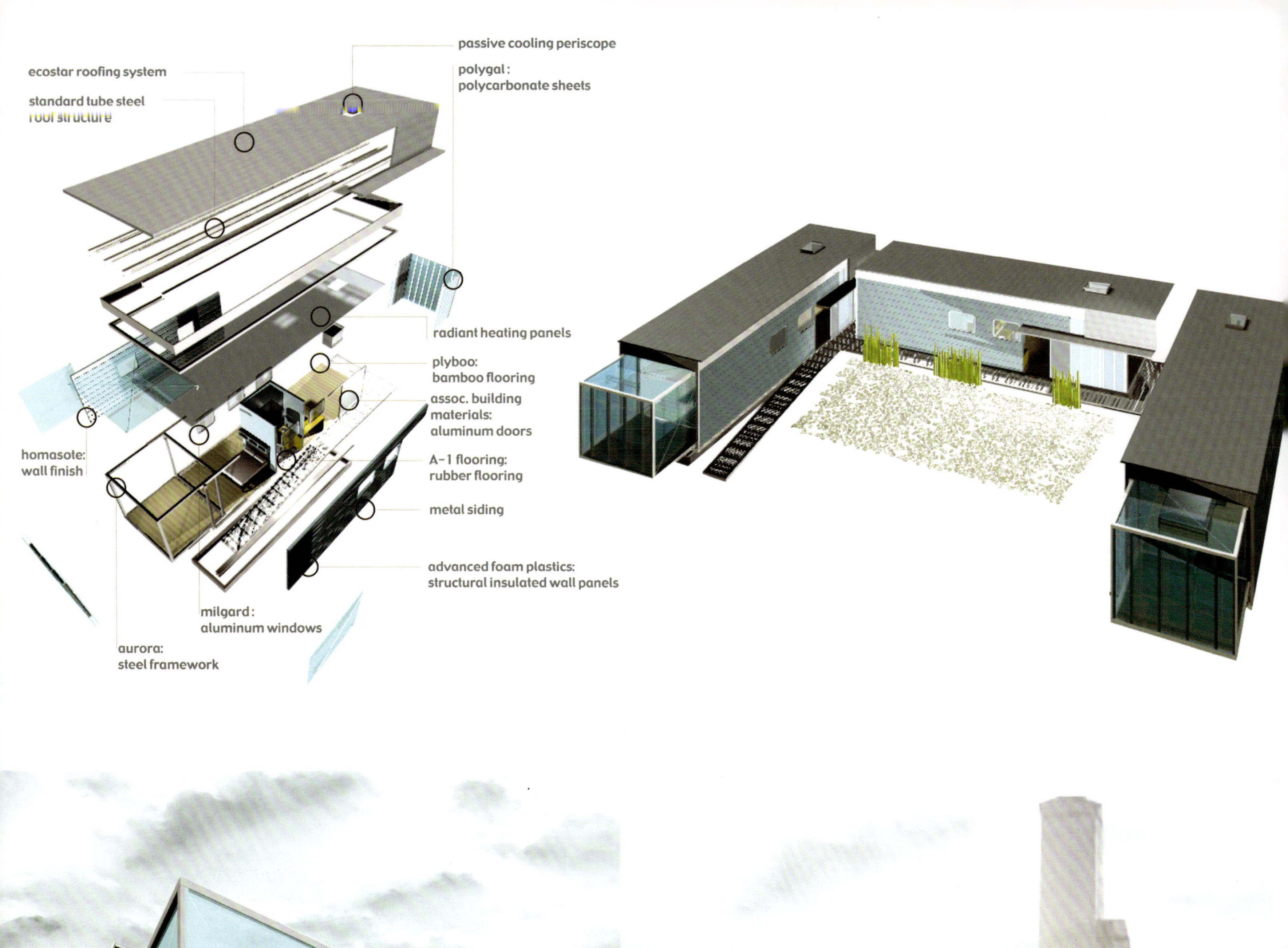

passive cooling periscope
ecostar roofing system
polygal:
polycarbonate sheets
standard tube steel
roof structure
radiant heating panels
plyboo:
bamboo flooring
assoc. building
materials:
aluminum doors
homasote:
wall finish
A-1 flooring:
rubber flooring
metal siding
advanced foam plastics:
structural insulated wall panels
milgard:
aluminum windows
aurora:
steel framework

siegalofficeofmobiledesign
omd

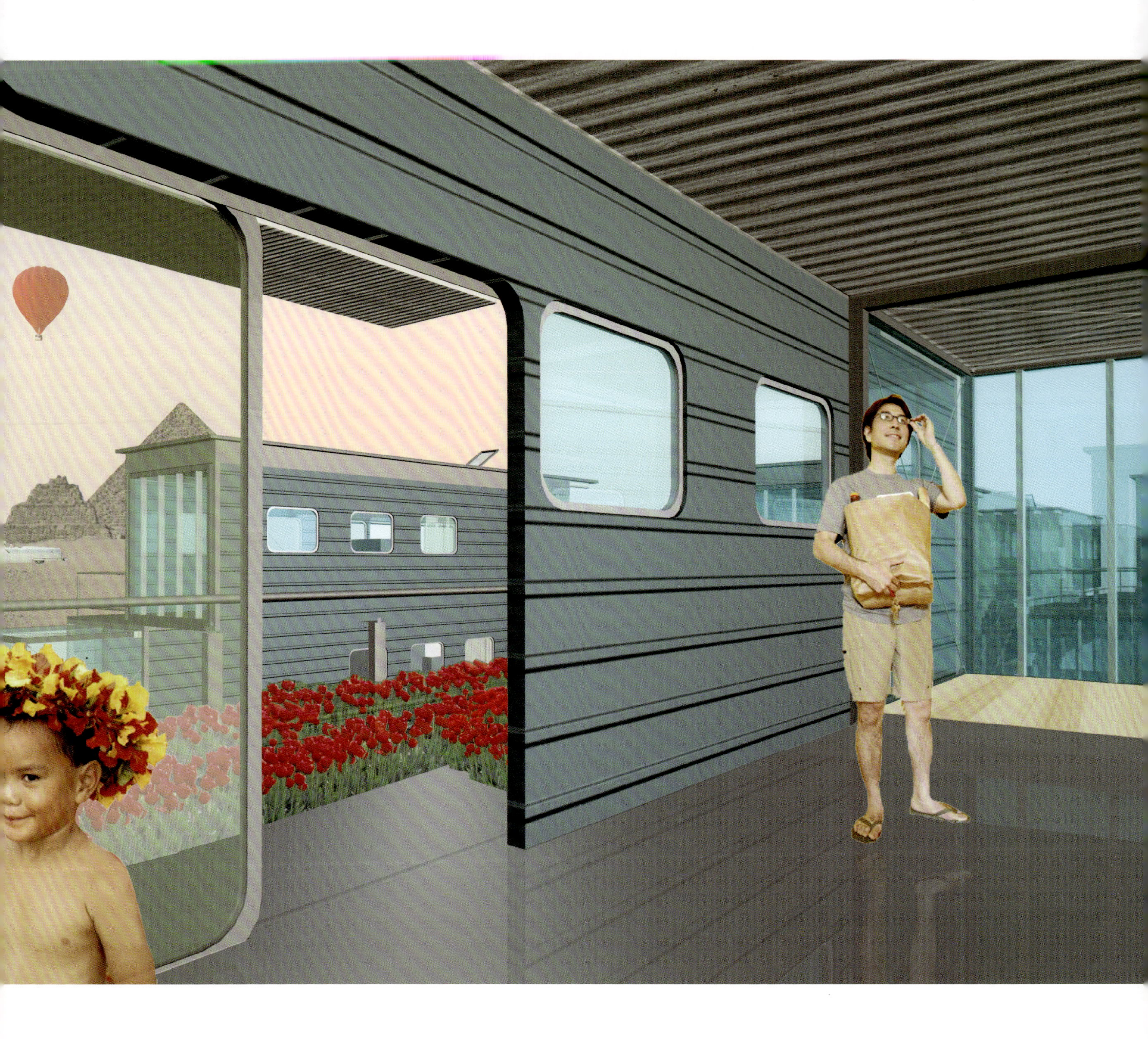

Above: rendering of Eco-Ville

生态村效果图

Contents

哈根达斯移动售货服务站

哈根达斯，2002年

哈根达斯的移动售货服务站让你在完全放松的同时体会奶油圆滑的质地，体验动感冰激凌吧/电影大屏幕的绝妙感觉。伸展的羽翼把冰激凌吧的自由形状从一个遮蔽的茧状物结合到广阔的空间中，自由地品尝爱。在这里，交流服务站的工作游刃有余，调酒师根据顾客需要调出热带芒果鸡尾酒、柠檬派雪糕或者巧克力午夜甜饼。移动售货亭通过移动售货和媒体体验的方式加强了哈根达斯品牌的视觉效果、认知度和独特性。在经济发展的基础上，形式跟随激情，这种适应性强的灵活结构总是能够及时地与瞬息万变的环境进行交流沟通。项目使用了与先进科技和结构结合的轻型材料。这个自给自足的自由式结构把哈根达斯各式各样的异域口味介绍给未来更多的人们。

Haagen-Dazs Pleasure Mobile

Completion Date: March 2002
Client: Haagen-Dazs
Principal: Jennifer Siegal
Design Team: Armando Hernandez, Marine Vanyan
Construction: Penwal Industries

The Haagen-Dazs Pleasure Mobile blends pure indulgence with creamy smooth texture. Experience the remarkable efficiency of the dynamic ice cream bar / film screening environment. Spread its wings and orchestrate the bar's metamorphosis from a sheltered cocoon to the space for free tasting love. It is here that the interactive service station effortlessly evolves, enabling mixologists to respond to clients' desires for Mango Tropicale Sorbet Cocktails, Lemon Passion Pie Licks, or Chocolate Midnight Cookie Bites. The Pleasure Mobile enhances the visibility, recognition and identity of the Haagen-Dazs brand through the mobile vending and media experience. Based on an economy of movement, where form follows passion, this adaptable and flexible structure is always responsive, interacting with its immediate and shifting environments. It is durably composed and constructed using lightweight technologically advanced, and structurally robust materials. This self-sufficient and relocatable structure introduces and reacquaints the diverse array of exotic Haagen-Dazs flavors to the expanding population in the metropolis of the future.

Left: structure view
结构

Right: the completed Haagen Dazs pleasure mobile

哈根达斯移动售货服务站

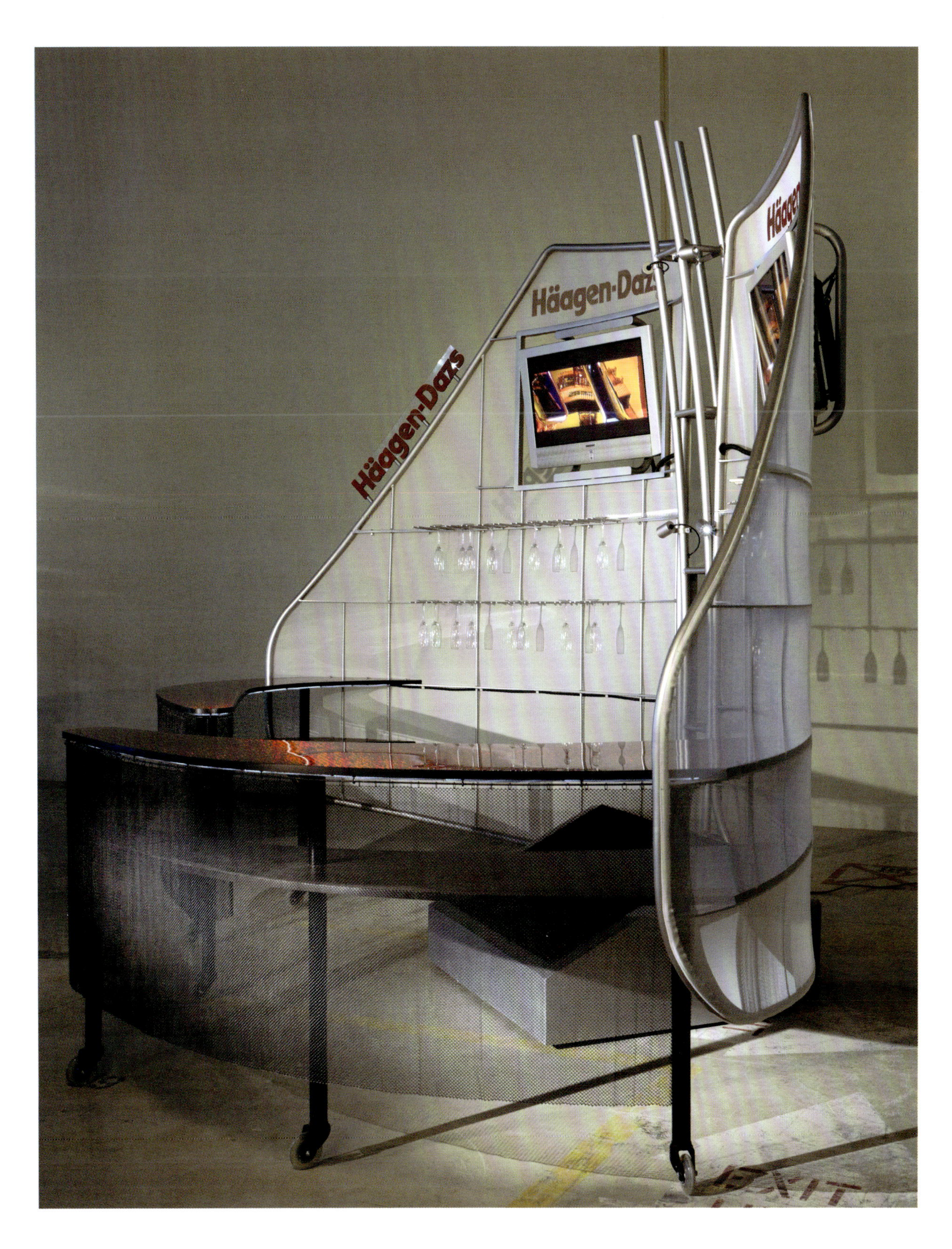

Opposite: the completed Haagen Dazs pleasure mobile

已建成的哈根达斯移动售货服务站

Häagen-Dazs

Left: rendering of the Haagen Dazs Pleasure mobile and structure photo

哈根达斯移动售货服务站效果图和结构图片

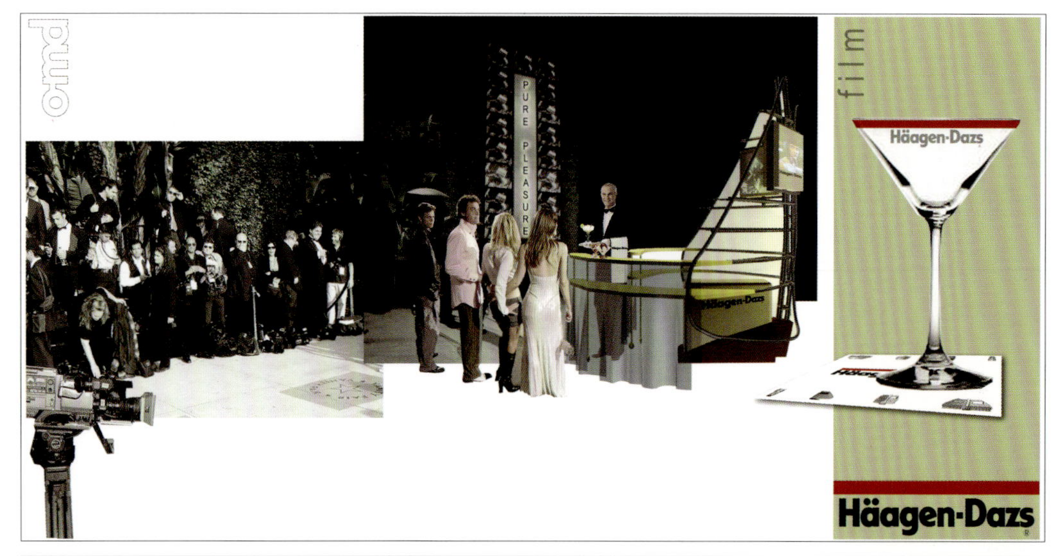

Below: completed Haagen Dazs Pleasure mobile
已建成的哈根达斯移动售货服务站

Häagen-Dazs

Contents

海德拉21

海德拉21是一个漂流救生建筑物，它为饱受战争痛苦的难民们提供了一个临时海洋避难所。组装是由直升机来进行，在海面上充气安装。外表使用的是一种由玻璃纤维加固过的合成橡胶，在保持充足采光的情况下使居住者不受外界-20～+120° F（约-29～+49° C）温度的侵害。当加上电压时，易弯的聚合物产生收缩，把海水抽吸到淡化海水的系统中，经过处理再放回到大海中。太阳能电池和一个利用海浪能源的系统为海德拉21提供了电力。一个漂浮的外部花园被固定在这个海洋住所的外面，为现代的海洋居住者提供了一个种植植物和休闲的好去处。

Hydra 21

Completion Date: Spring 2004
Client: Popular Science Invited Competition
Principal: Jennifer Siegal
Design Team: Lena Schacherer,Carina bien-Willner

The buoyant survival structure Hydra21 provides a temporary ocean refuge for citizens of war-torn nations. Distributed by helicopter, the prefabricated units inflate on impact with the ocean surface. The structure's outer skin, a synthetic rubber reinforced with a glass-lattice fabric, protects the inhabitants from temperature extremes ranging from -20°F to +120°F and lets in ample light. Pliant polymers that flex when a voltage is applied pump seawater to the desalinization system; treated wastewater returns to the ocean. Solar cells and a system that harnesses wave energy supply the Hydra 21 with electricity. A floating exterior garden, anchored to the marine dwelling's outer skin, gives modern-day ocean dwellers a place to grow food and to stretch their legs.

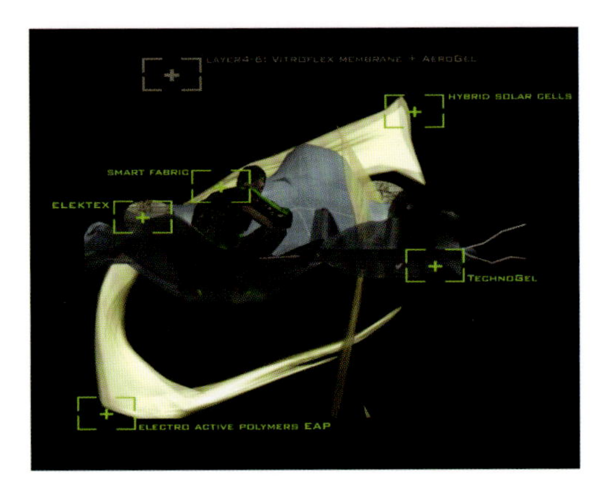

Opposite: rendering of Hydra 21
海德拉 21 效果图

Left: inspirational image
灵感来源

care

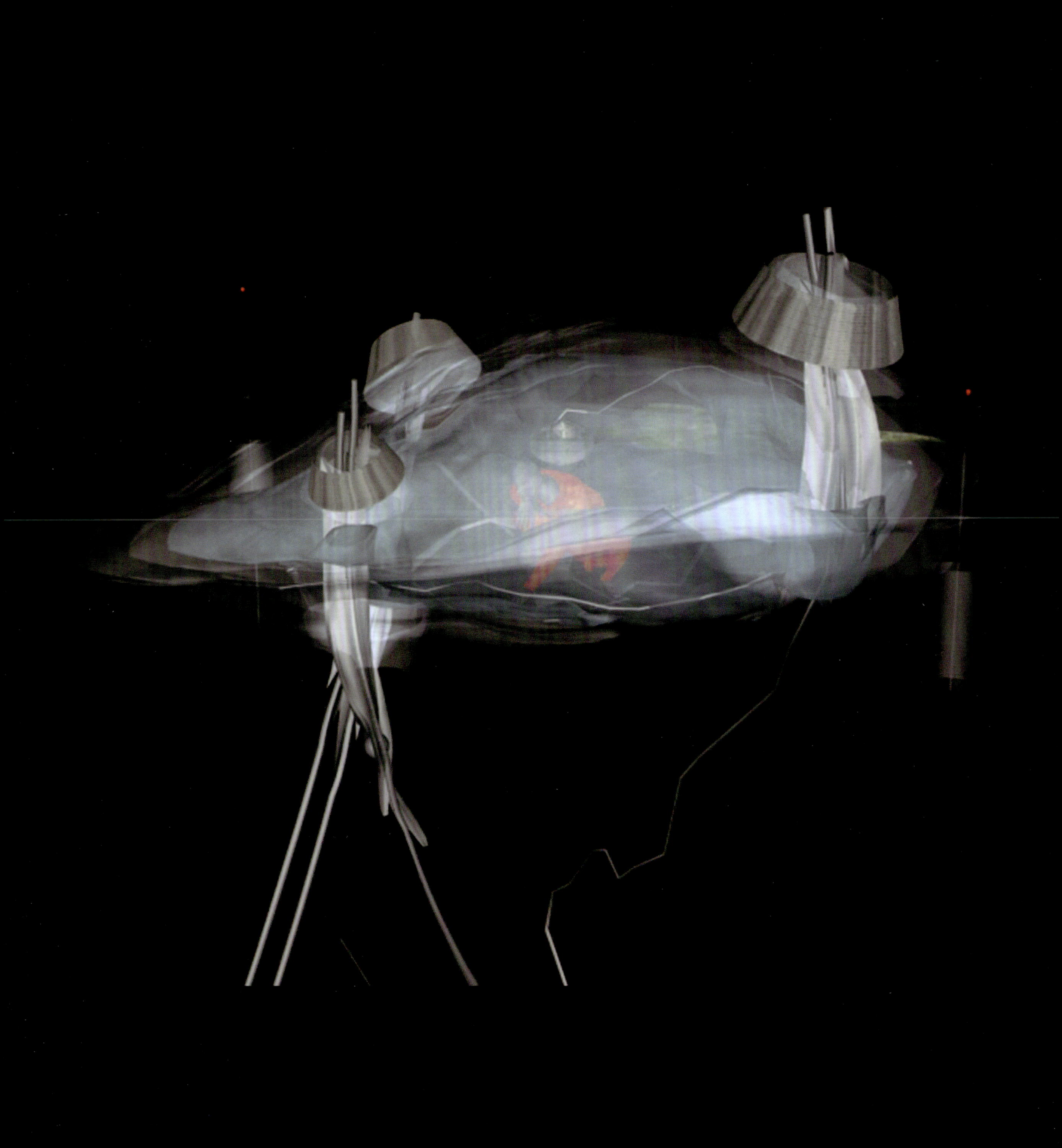

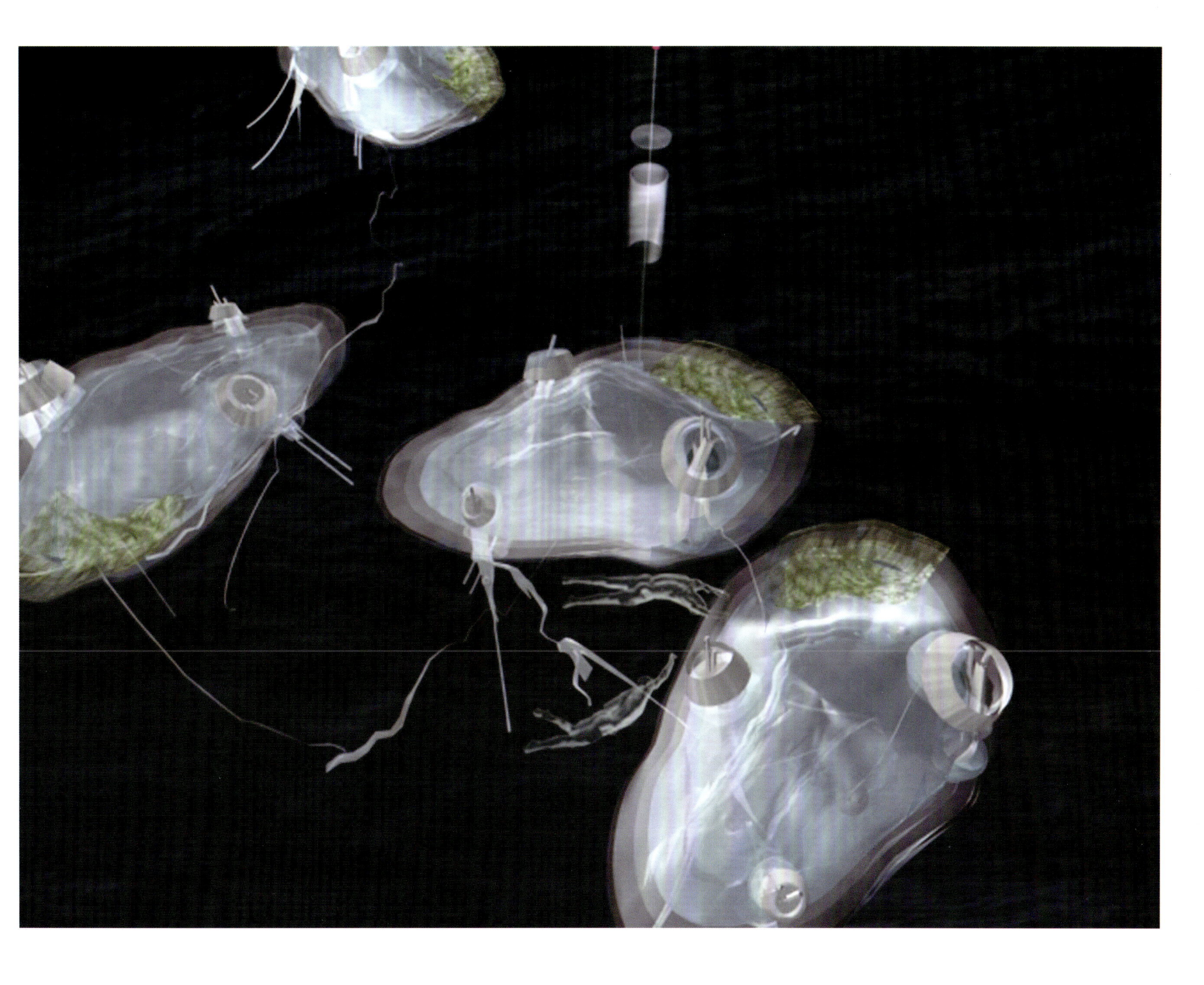

Opposite and previous pages:
rendering of the Hydra 21
海德拉 21 效果图

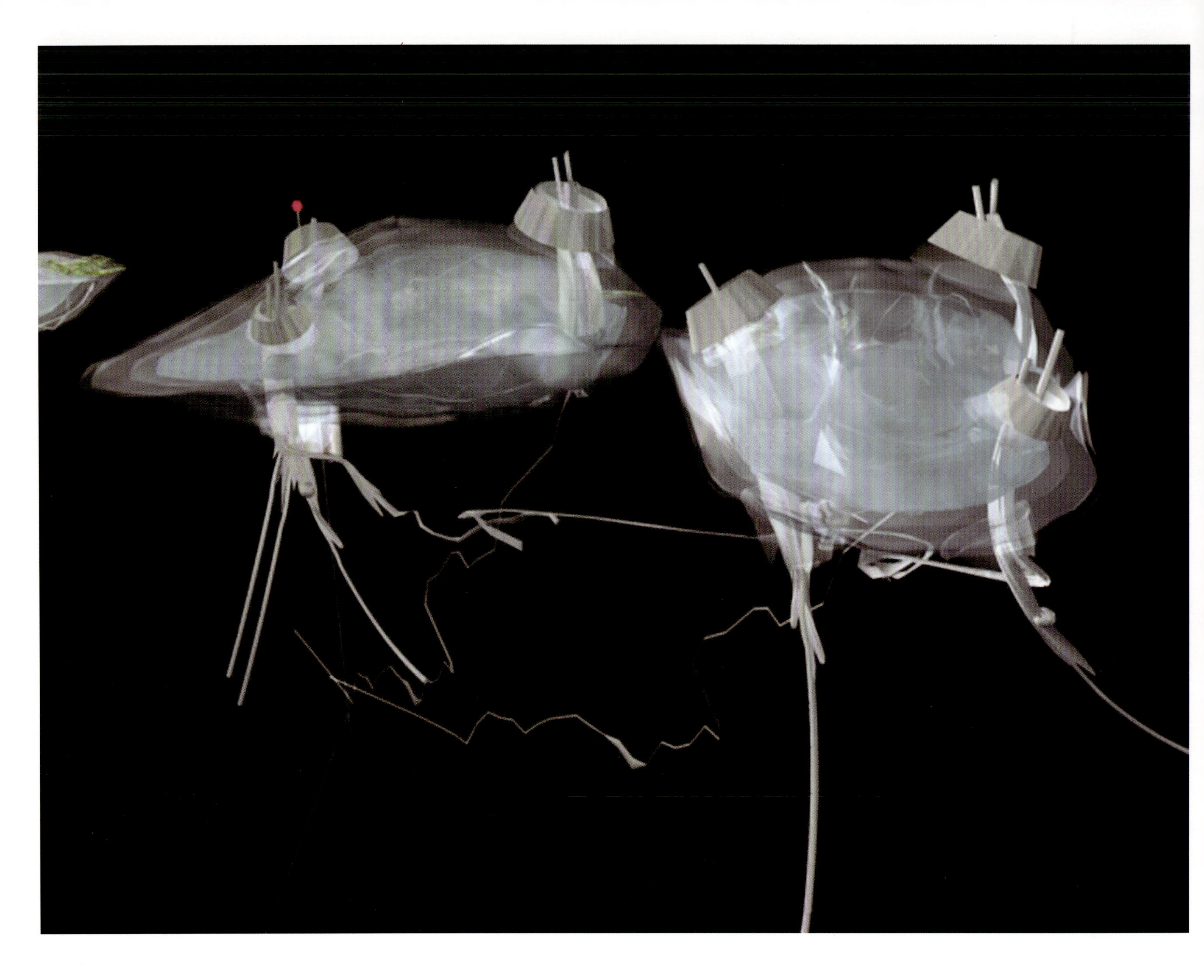

Above: rendering of the Hydra 21

海德拉 21 效果图

Contents

海德拉住宅

海德拉住宅是一个针对全球气候变暖、海水脱盐和循环问题而设计的大型的移动定制建筑。

整个建筑结构体分为底架（提供内部结构、动力、通信、机械和自供能源收集系统）、组装元素（为内部扩建、内外部表面、电子和通信）。这是个由部件设计、工程技术和系统层面共组的一体化设施。设计的重点在于独特的规划和建筑环境，并引导个体居住者根据他们的需要和价值观来适应居住环境。这样一来，与一次性组装结构有关的额外成本、风险和协调错误都被避免了。

结构的智慧体现：

1.水：装满雨水的气囊、脱盐水（地球上的水有97%都是海洋里的盐水）和废水。每根管道都能将海水抽升（底部直接从大洋里抽水），再向下分配脱盐后的水以提供饮用和盥洗。
2.动力：光电技术、水盐结晶技术和热能导体技术。
3.通信和机械：全球资讯和管道设备

气动外部表面：2层充气氯化橡胶
液化连接：每一个独立的单元都安装有像触角一样的压力泵，并形成一个外部通道群。

悬浮花园：每一个独立的单元都设有各自的悬浮花园。这些像睡莲浮叶的花园环绕于海德拉住宅的周围；与建筑主干相连的脐带为花园输送新鲜水分和养料。

Hydra House

Completion Date: January 2003
Client: Wallpaper Magazine Invited Proposal
Principal: Jennifer Siegal
Senior Design Associate: Kelly Bair
Assistants: Sara Schuster, Peter Klein

hy.dra: n. Any of several small freshwater polyps of the genus Hyrda and the related genera, having naked cylindrical body and an oral opening surrounded by tentacles.

Hy.dra: n. The many headed serpent that was slain by Hercules.

hy.dranth: n. A feeding zooid in a hydroid colony, having an oral opening surrounded by tentacles. (hydr(a) + Gk. anthos, flower.)

The Hydra House is a mass-customized mobile modular structure that is responsive to environmental issues of global warming and water desalination and recycling.

The structural stalks are separated into chassis (providing internal structure, power, communication, mechanical, and a self-sufficient energy collecting system), and mass-customized elements (for interior build-outs, exterior and interior skins, electronics, and communications). Component design, engineering, and integration are at the system level. This allows architects to concentrate primarily on the unique programmatic and environmental context of the building, and allows individual occupants to focus on tailoring their environment according to needs and values. In doing so, the additional cost, risk, and coordination errors associated with a "one-off" or mass-customized structure are avoided.

Structures, embedded with intelligence:
1. Water: rain water with stretched bladder and desalination (97% of the planet's water is salt water in the seas and oceans) and treated waste water. Each tube either pulls sea water upward (see bottom skin punctures drawing directly from the ocean) or distributes desalinized water

Above: inspirational image
灵感来源

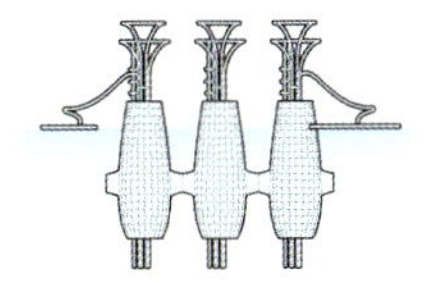

Below: rendering of Hydra House
海德拉住宅效果图

downward to provide potable and washing water.

2. Power: photovoltaics, salt crystalization, and thermocouple energy conductors

3. Communication + Mechanical: global knowledge and plumbing

Pneumatic Exterior Skin: 2 layers of inflated neoprene

Liquefied Connections: suction-like tentacles attach to each independent housing unit, forming colonies and allowing for external passage.

Floating Garden: each independent housing unit has an attached self-sufficient floating garden. These Lily pads stem from Hydra House's structural stalks, using an umbilical cord to provide fresh water and nutrients gives life and feeds the floating garden.

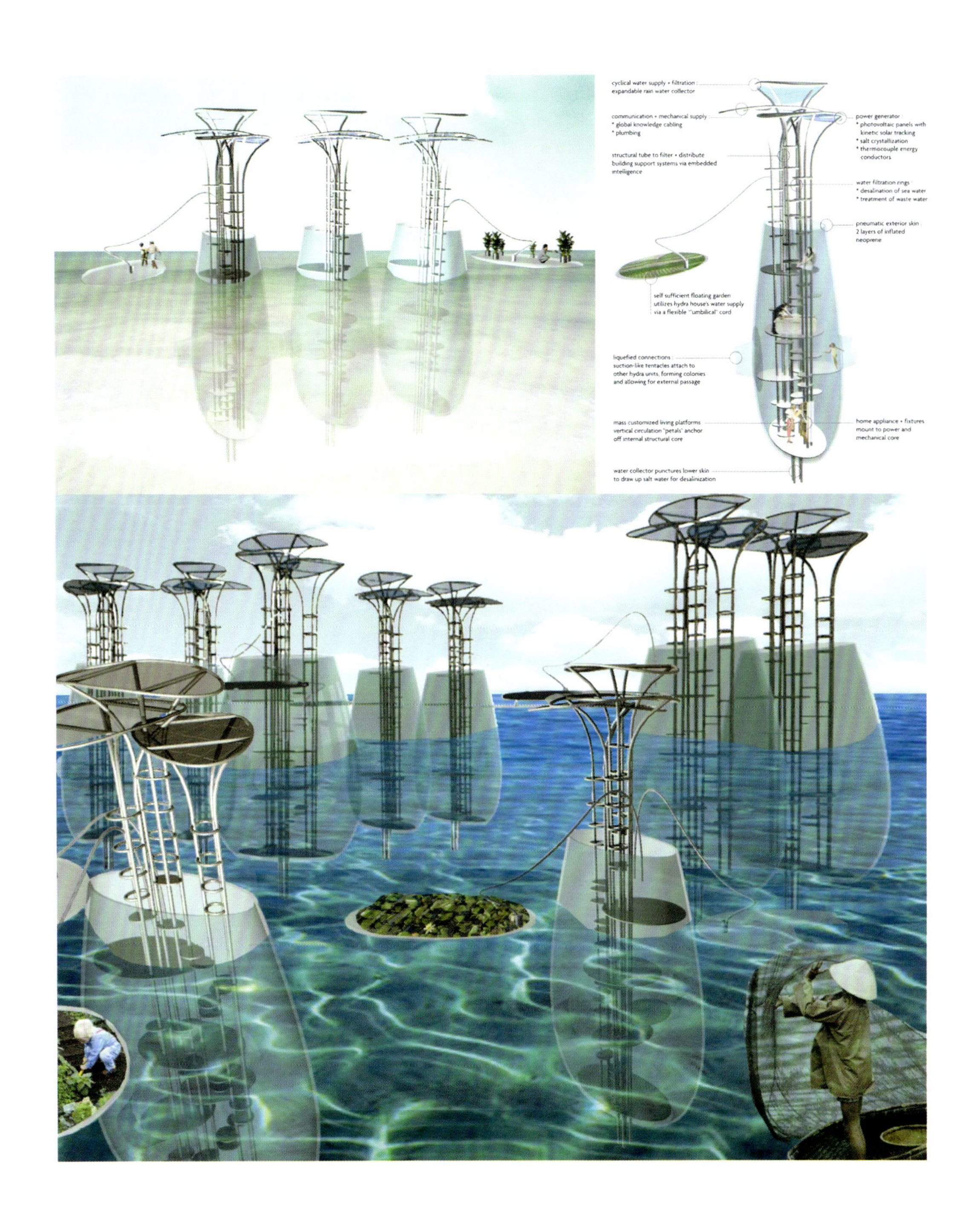

Contents

移动城建筑

移动城是提供给社会活动策划公司进行关于艾滋病、乳腺癌和饥饿等问题的多日户外慈善活动使用的，主要目标是为人们提供便于晚间露营的拆装简易的结构，并能适应不同场地和气候条件。它干净、明亮和醒目的外观是社会交流的亲密媒介。

露营地的构件被预设为四个不同层面，既可以围绕一个中心聚集空间（中心广场模式）搭建，也可以呈线性的廊道排列（主大街模式），或者互相混合在一起。

对一些个体的露营地元素，诸如医疗服务厅、市场、售货亭和“追思堂”等设施的设计是以现有的敞篷货车类型为基点，利用张拉构造结构把它们结合起来。这些紧凑的、自控的移动结构设施能够适应天气的变化，卫生而且便于拆装。各部件在伸展、滑动和推拉的基础上能够摊开地板，搭起屋顶、墙壁和悬垂物。这使得原有的机车变成了独特而美妙的建筑机器。

规划要求基本建筑结构位于一个抬升的木板路上，既可以沿着“主大街”延伸，也可以围绕“中心广场”安置。这条抬升的通道从不同的机车上滑拉下来连接而成，并把移动城中的各种不同元素合为一体，它还避免了因地面的不平坦而造成的流通上的困难。这一切都使得露营地的活动操作更加便捷，令活动参加者在各个方面感到舒适方便。

Mobile Event City Architecture

Completion Date: winter 2001
Client: Pallotta TeamWorks
Dan Pallotta, Founder/Chief Executive Officer
Principal: Jennifer Siegal
Design Team: Greg Roth,Elmer Barco,Thao Nguyen,Jon Racek

The MOBILE EVENT CITY provides an overall upgrade of event facilities for a socially conscious event-planning firm that creates multi-day outdoor charity events in support of AIDS, breast cancer, world hunger, and like causes. The goal is to provide structures for their nightly encampments that are easily relocatable, adaptable to varying site conditions, climate controlled, clean, well-lit, and visually striking - and that help engender intimate social interaction.

Four possible master plan schemes assemble the campsite components into a four-tiered hierarchy, which can then be organized either around a central gathering space ('Town Square'), along a linear corridor ('Main Street'), or in a combination thereof.

Individual campsite elements, such as Medical Services, Outreach (marketing), Vending Kiosks, and the 'Remembrance Place,' use existing truck types as points of departure, then hybridize them with tensile fabric structures. These compact, self-contained mobile structures are weather resistant, hygienic, and easy to deploy and relocate. As they unfold, slide open, pivot and pull apart to expand their floor areas, their fabric components take shape to form roofs, walls and overhangs, transforming their host vehicles into unique, wondrous building/machines.

The master plans call for the primary structures to be situated along an elevated boardwalk that either extends linearly along 'Main Street,' or circumscribes the 'Town Square.' This raised thoroughfare, composed of sections that slide out of each vehicle and interlock, unifies the disparate elements to the Mobile City, providing identifiable and accessible circulation that is also a level alternative to an often uneven ground plane. This streamlines the organization of campsite operations, and augments the overall level of comfort and care afforded event participants.

Opposite: rendering of mobile event city
移动城建筑效果图

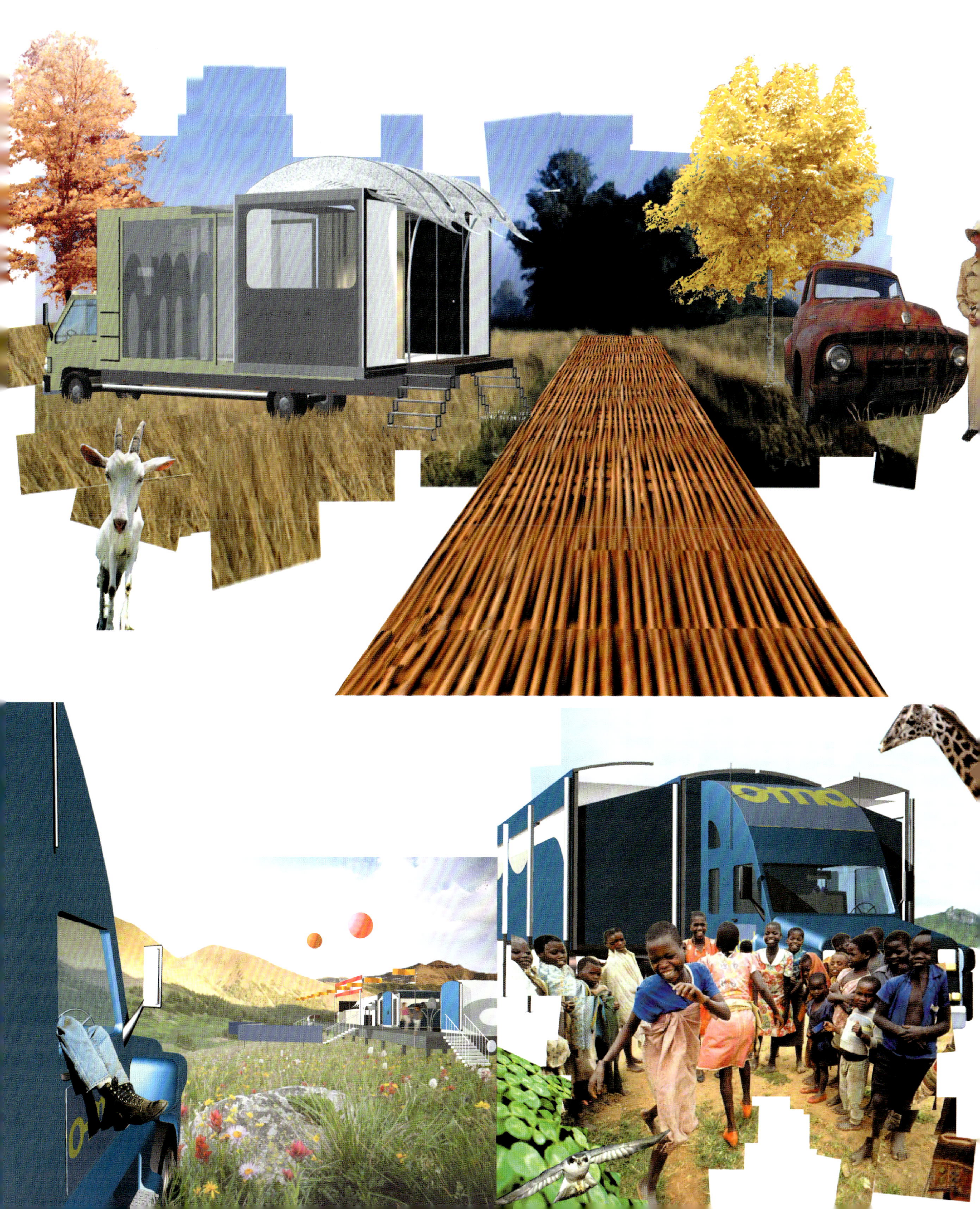
omo

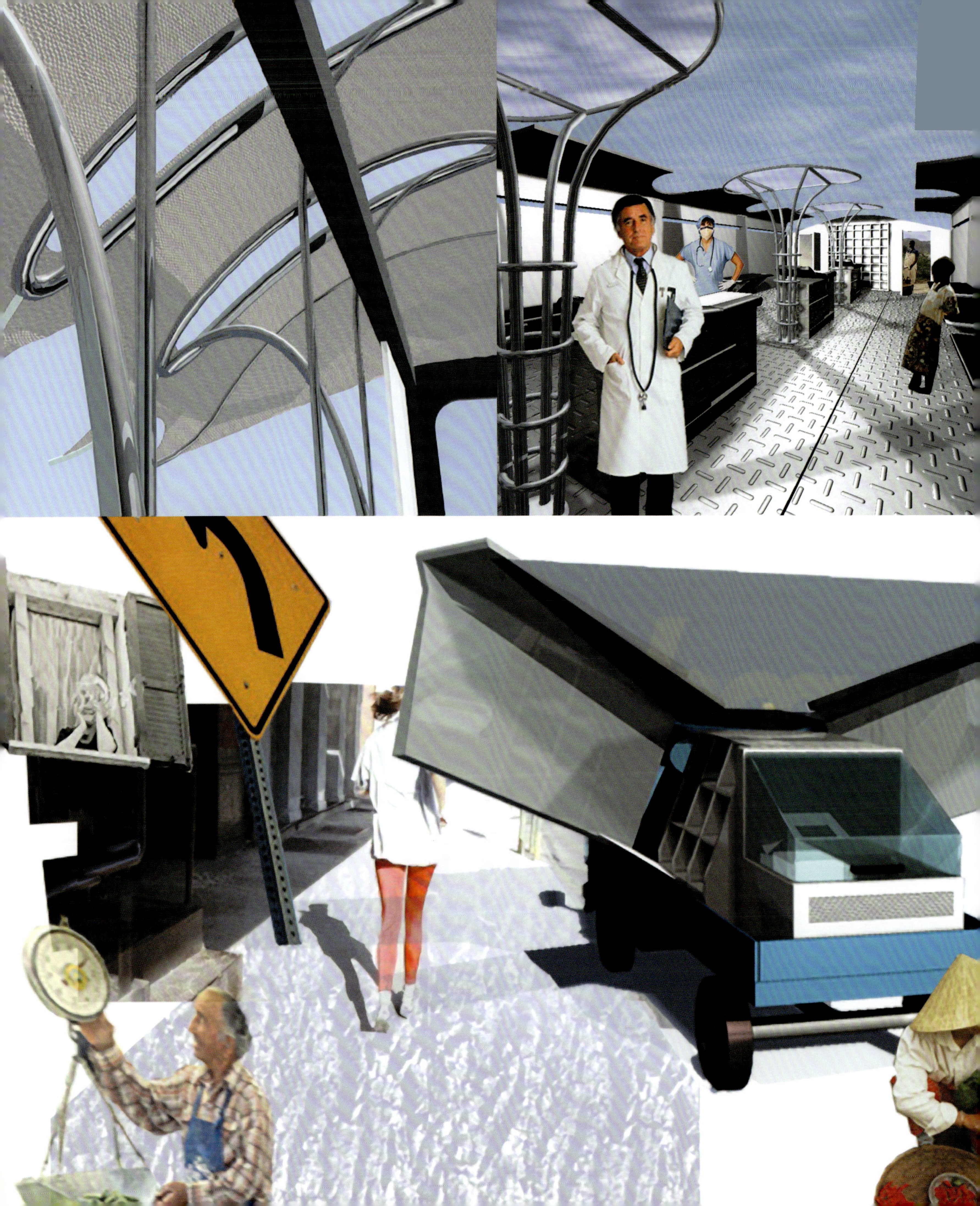

Contents

iMobile

iMobile是一个能够进入全球通信网络和通告最新计算机系统的外围设备和硬件、软件的在线流动端口。

这是一个充满活力的移动体，拥有惊人的高效率。工作场所中的一切活动都开展得有条不紊，能够适应业务的增加、减少和变化的需要。iMobile为移动企业家提供了一个建筑解决方案。在形式服从需要的经济发展背景下，这个灵活的适应性强的建筑物总是能对周围瞬息万变的环境作出反应。这个自供和移动的结构由高质量、轻质和经济型材料建成，它为人类的未来定义了一种新的都市形态。

iMobile

Completion Date: Research and Design completed Spring 2000
Client: OMD Research Projects
Principal: Jennifer Siegal
Design Team: Elmer Barco,Arona Witte
Project Team: Ashley Moore,Saul Diaz,Jason Panneton

Fold-out. Plug-in. Boot-up.
The iMobile is an online roving port for accessing the global communications networks and announcing the latest computer systems, peripherals, hardware and software.

Marvel at the remarkable efficiency of a dynamic mobile enterprise. It is here that the workplace effortlessly evolves, enabling businesses to respond to the need for augmentation, contraction and metamorphosis. The iMobile offers building solutions to the mobile entrepreneur. Based on an economy of movement, where form follows necessity, this adaptable and flexible structure is always responsive to its immediate and shifting environment. Composed and durably constructed from high-quality, light and affordable materials, this self-sufficient and relocatable structure gives shape to the metropolis of the future.

Left and opposite: exterior and interior view of the imobile
imobile 模型内外景观

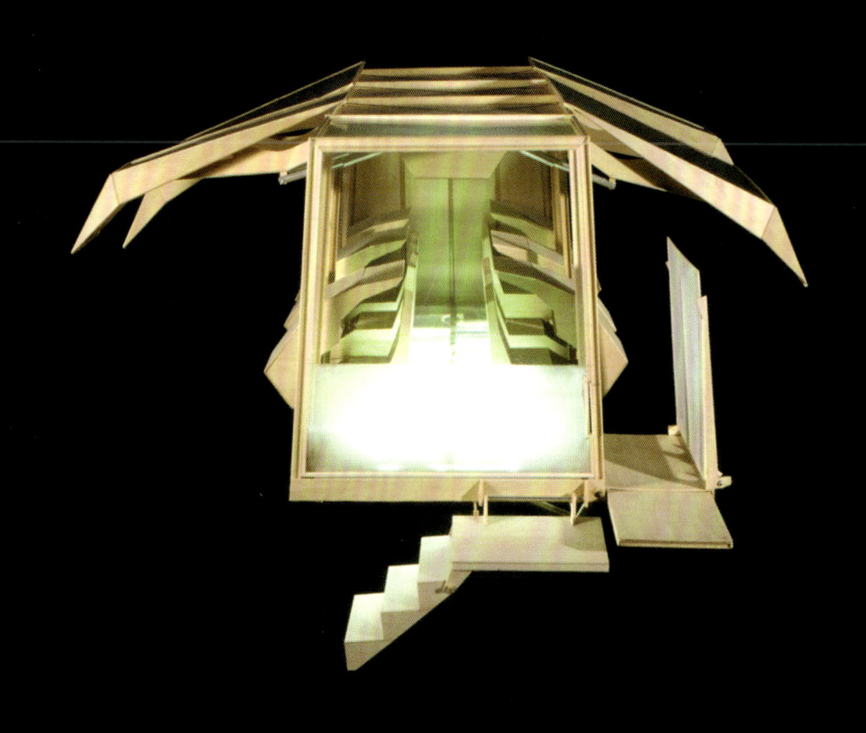

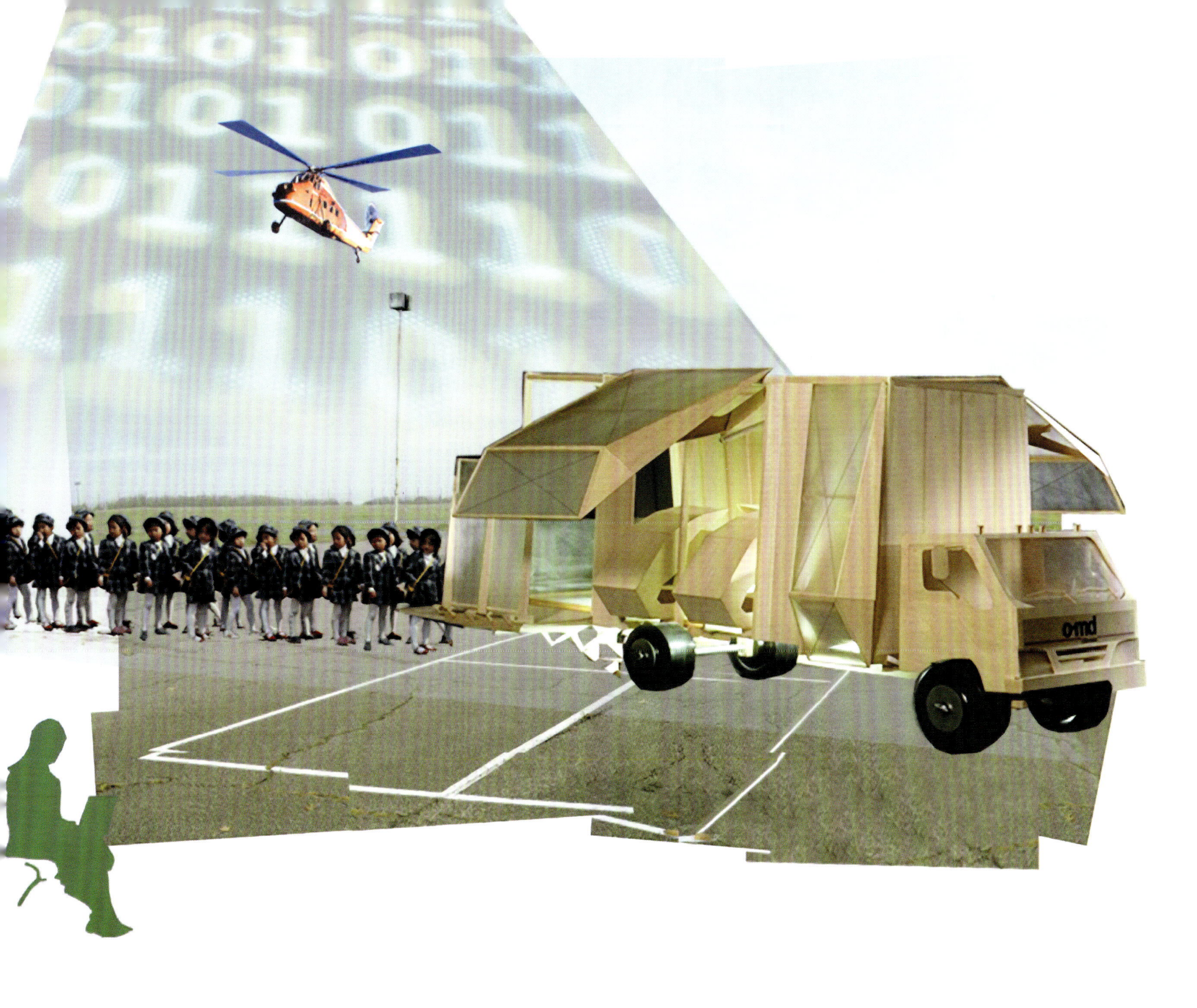

Opposite and above: rendering and model of imobile

imobile 模型和效果图

料
公
司
蟹
辣
iMobile
4
3
1
o-md
991119

Contents

移动构造训练中心

威尼斯，加利福尼亚州

移动构造训练中心是为威尼斯社区住宅公司——一个旨在为残疾人和低收入人群开发和维护经济型住房的组织所设计的移动学习设施。这个非营利组织给予他们的培训者学习建筑技术的机会。作为回报，学生们再把他们所学到的技术运用到所需的项目中。 移动构造训练中心的面积为14英尺× 65英尺，这个实践性教室为四种建筑构造的基本技术：管道技术、粉刷和石膏修复、木工和电工技术提供学习空间。

这个设计理念鼓励学生和老师之间的视觉交流。中心的入口处有一个面积14英尺×14英尺的会议室，这里用于陈列建筑构造范例板，同时也是一个聚集了不同建筑形式的交流空间。拖车叠起来就显现出一个像通廊似的大型室内工作站，便于教师进行流动的督查和交流。

在90° 位置上的可操作透明板不但具有遮蔽功能，而且能调节冷热空气。 另外，移动构造训练中心的尽头连着一个木材商店，这里是一个超越房车之外的自由实践基地。

移动、灵活和可操作的特点使得移动构造训练中心成为教授可选择性建筑技术的样板教室。

Portable Construction Training Center (PCTC)

Completion Date: August 1998
Owner: Venice Community Housing Corporation, Steve Clare, Director
Principals: Jennifer Siegal,Lawrence Scarpa
Woodbury University Student Project Team: Alex Arias, Thomas Cohen, Guillermo Delgadillo, Maurice Ghattas, Han Hoang,Chayanon Jomvinya, Thao Nguyen, Jose Olmos, James Popp, Phung Thong, Juan Uehara
Assistants: Wendy Bone, Robert Chambliss Ezell Edmond, Ann Murphy, Gwynne Pugh

The Portable Construction Training Center was conceived for the Venice Community Housing Corporation, an organization founded with the mission to develop and maintain permanently affordable housing for disadvantaged and low-income individuals. This non-profit organization affords an opportunity for their student trainees to learn construction skills and in turn apply their skills to needed projects. The 14 x 65 foot PCTC is a hands-on classroom used as the focal point in this construction training process. The PCTC allows space for the four basic construction trades: plumbing, painting and plaster repair, carpentry, and electrical.

Right: inspirational image
灵感来源

Above: digital rendering of the concept design for PCTC, a view of the porch side

移动构造训练中心效果图，走廊一侧

Upper: rendering of PCTC and the completed PCTC on road
Bottom: construction sequence and wood shop view

移动构造训练中心效果图和走在路上的完成建筑
施工现场

The design concept encourages visual connections between apprentice and teacher. There is a 14 × 14 foot meeting space at the PCTC threshold that exhibits construction example boards, and provides a well-lit location to gather between building sessions. Like a large porch, one entire length of the trailer folds open to reveal interior workstations. This creates a catwalk used by the teachers to facilitates inspection and interaction.

In this 90 degree position, the operable translucent panels give shade and regulate the natural flow of hot and cool air. Additionally, the far end of the PCTC folds open revealing a wood shop, where tools can be disengaged and utilized beyond the parameter of the trailer.

Portable, flexible, and operable, the PCTC is a symbol for alternative construction techniques and provides a place to teach them.

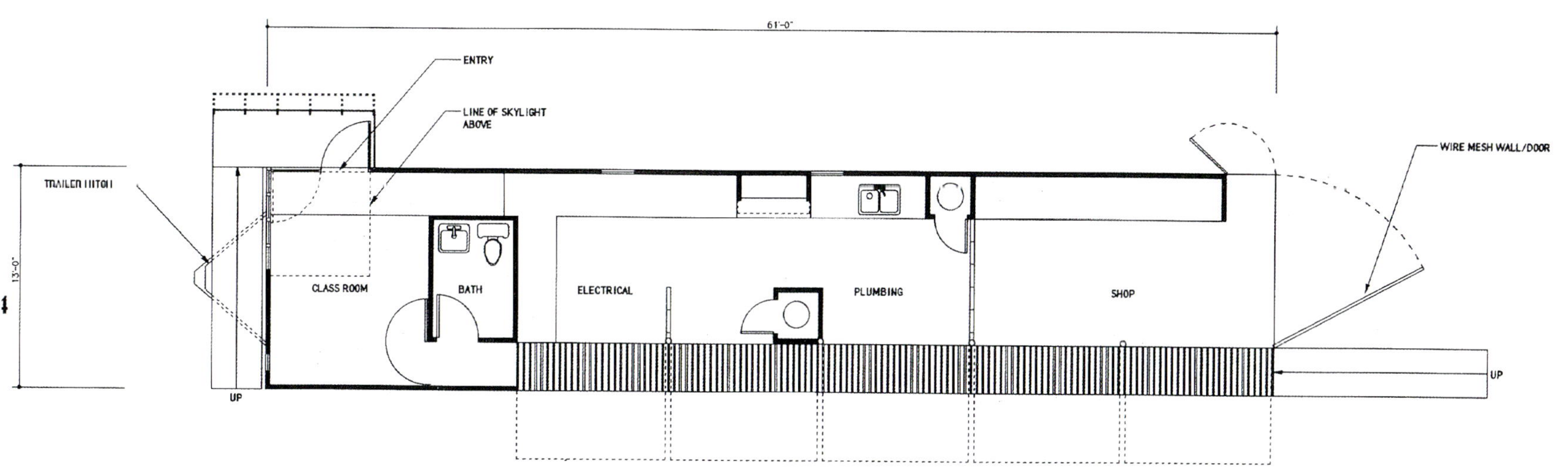

Upper left: dusk view of the completed PCTC

黄昏中已建成的移动构造训练中心

Upper right: porch side view of the completed PCTC

走廊一侧

Bottom: drawing of the concept design

概念设计图

Above: interior view of the completed PCTC

移动构造训练中心室内

Below: exterior view of the completed PCTC

移动构造训练中心外观

WILD JOHNNY'S
WILD JOHNNY'S
WAGON
CATERING
884-1558
WILD JOHNNY'S

Contents

轻便住宅

回忆过去的轻便式庇护所和住宅的模式，如今的轻便式住宅在不断变化的环境中改写了自身的价值和形象。 对于如今日趋昂贵的房价，它在为我们提供了一个环保而经济的选择的同时， 也引发了关于巡回式家园和公园的新的设想，对那些收入颇丰，却无法进入因缺乏资源而萎缩的传统住宅市场的人提供了全新的选择。

轻便式住宅空间的可延伸和可收缩性，其材料的不同透明度和可移动性体现了其独特的灵活和适应性强的特点。以中心厨房/浴室为核心，把卧室区同餐厅/起居间划分开，各部分的形式和功能设置都很紧凑。当需要额外的空间时，起居室可以外扩以求得新的空间。 设计使得这种住宅可以根据需要来进行调整和重新定向以获得满意的自然光和新鲜的空气。

作为一种存在于自身的实体， 轻便式住宅每到一处都能够适应并创造新的社会形态。当个体单元被聚集在一起时就形成社会交流的公共空间，比如花园、庭院等等。一个复合的空间形态也能被单个业主拆分，并与相邻的生活、工作和社交空间产生新的连接形式。

轻便式住宅的可移动性和在环境中的易安置性同其他房屋形成了鲜明的对照。 它提醒我们当外部环境动荡不安时，这是一种多么容易为人类提供庇护的住宅元素。轻便式住宅为每天的生活带来了灵活多变的社会形态。无论是在野外风景中做一个短暂的停留，还是偶尔出现在都市空间里，或是做一个长久的停留，功能齐备的它总是能够满足多种需要。

Portable House

Completion Date: under construction
Clients: Dr. Lance Stone
Principal: Jennifer Siegal
Design Team: Kelly Bair, Elmer Barco, Thao Nguyen, Jon Racek, Andrew Todd

Harkening back to original prehistoric models of shelter and dwelling, the Portable House adapts, relocates and reorients itself to accommodate an ever-changing environment. It offers an eco-sensitive and economical alternative to the increasingly expensive permanent structures that constitute most of today's housing options. At the same time, the Portable House calls into question preconceived notions of the trailer home and trailer park, creating an entirely new option for those with disposable income but insufficient resources for entering the conventional housing market.

The Portable House's expandable/contractible spaces, the varying degrees of translucency of its materials, and its very portability render it uniquely flexible and adaptable. Its central kitchen/bath core divides and separates the sleeping space from the eating/living space in a compact assemblage of form and function. When additional space is required, the living room structure can be extended outward to increase square footage. By design, the House can be maneuvered and reoriented to take advantage of natural light and airflow.

As an entity unto itself, the Portable House adapts to or creates new social dynamics wherever it goes. For example, when individually owned units are grouped together, they can create common spaces for social interactions, such as gardens, courtyards, side yards, etc. Or multiple units can be arranged by one owner to create separate but adjacent spaces for living, working, and socializing.

The Portable House's mobility, the way it moves across and rests lightly upon the landscape, provides a provocative counterpoint to the status quo housing model. It recalls a time when the elements that constituted shelter were easily manipulated to accommodate innumerable variables and conditions. It likewise offers flexibility in the socio-dynamics of everyday living. Whether momentarily located in the open landscape, briefly situated in an urban space, or positioned for a more lengthy stay, the Portable House accommodates a wide range of needs and functions.

Below: rendering of the portable house

轻便住宅效果图

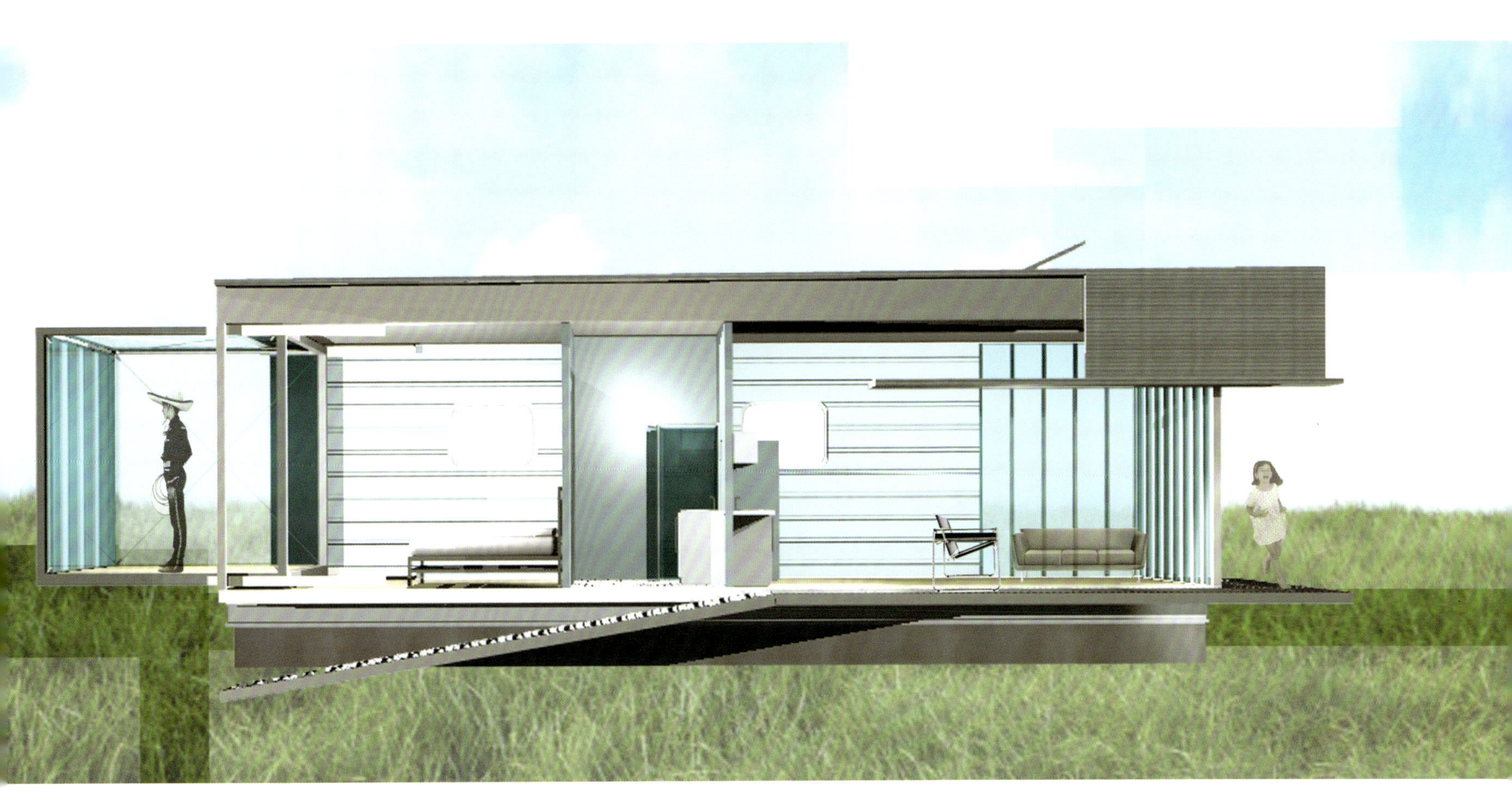

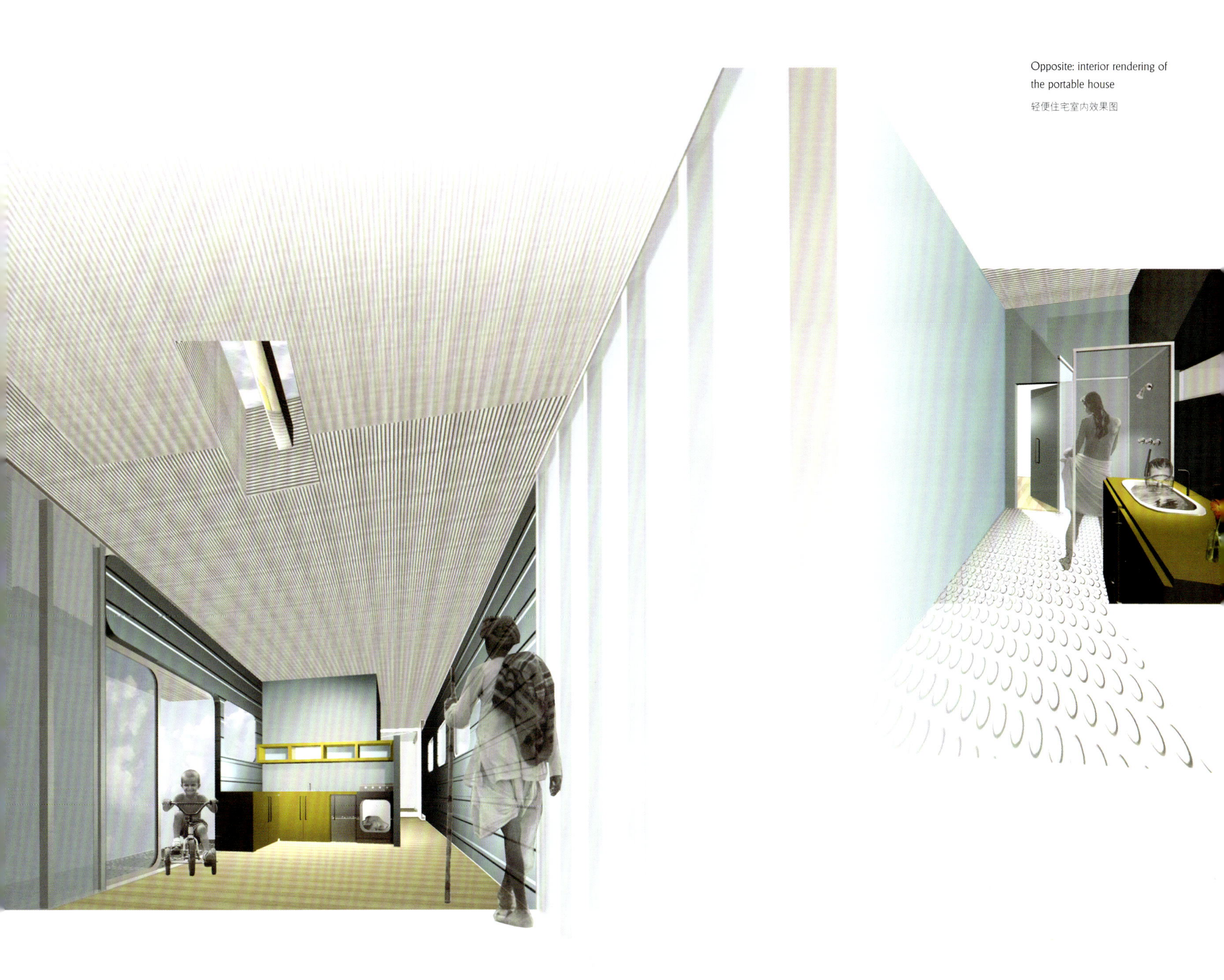

Opposite: interior rendering of the portable house

轻便住宅室内效果图

Above and below: rendering of the portable house

轻便住宅效果图

Opposite and next page: modeling of the portable house

轻便住宅模型

Contents

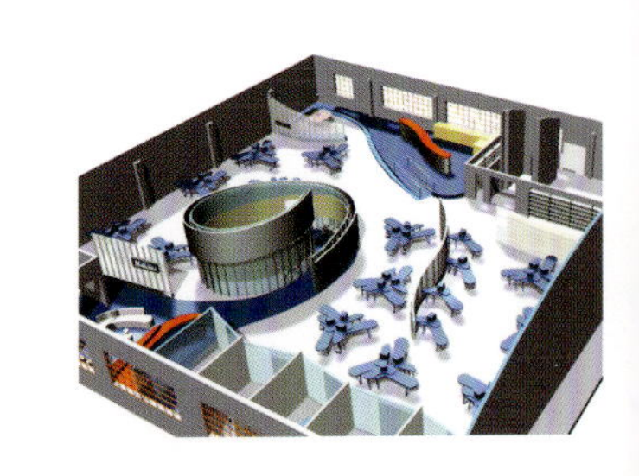

PIE.com Corporate Headquarters

Completion Date: October 2000
Client: PIE.com
Sebastian Copeland, Co-Founder and Chief Creative Officer
Principal: Jennifer Siegal
Design Team: Naoto Sekiguchi, Elmer Barco, Thao Nguyen, Greg Roth
Assistants: Ariana Rinderknecht Ashley Moore
Contractor: Crommie Construction- Jim Iwanski
Structural Engineer: Robert Englekirk- Lawrence Ho
Electrical Engineer: Antieri & Associates- Rolono Fortello
Steel Fabrication: Robert Chambliss
Work Station Fabrication: Jaime Ramirez
Reception Desk Fabrication: Robert Chambliss
Artist: Andre Miripolsky
Color Consultant: Emile Keff

For its corporate headquarters, the extreme sports-oriented web site pie.com chose a bow truss loft space in Hollywood. The open plan of the 10,000 s.f. space houses the various functions of the business under one roof, and allows for a dynamic and free flow of [ideas, information, views, light, bodies] within that space. The program includes: reception lobby, conference room, work stations, executive offices, kitchen/dining, storage/copyroom, and brainstorming 'base camp' space.

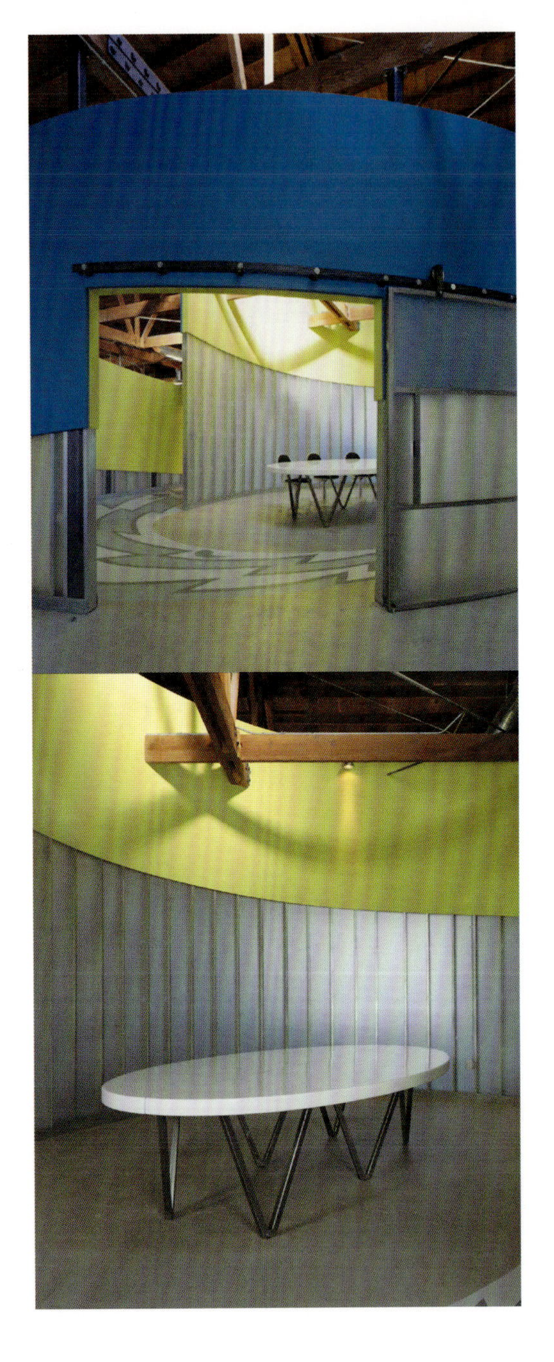

PIE.com公司总部

极限运动网站pie.com为它的总部所在地选择了好莱坞一个有着穹顶结构的阁楼空间。这个占地10 000平方英尺的建筑空间为在同一个屋檐下进行的不同商业活动提供办公场所，这个空间引导着思维、信息、景象、光线和实体的自由流动。整个项目包括：接待大厅、会议室、工作站、行政办公室、厨房/餐厅、储藏室/复印室和“头脑风暴法”学习营地。

一系列随处可见的轻巧、精致的结构以曲线玲珑的和半透明的形式蜿蜒曲折地划分了开会、休息、吃饭和工作的空间。室内工作区点缀着许多三个一组的围绕着电源供应柱的单色阿米巴虫形桌子。这些工作区的位置由头顶的环绕整个办公室的电线路径所决定。一间很大的螺旋形会议室支配着整个空间，像升向天空的颤动着的蓝绿色海浪。从电梯口沿着弯曲的灰色地板［由加州波普艺术家

Above and right: pie.com loft space view

pie.com 阁楼空间

To distinguish work areas, a series of light, almost delicate structures appears throughout, curvaceous and translucent, snaking here and there to delineate places for meeting, sitting, eating, and working. Dotting this interior work-scape are clusters of bi-level, amoeba-like desks gathered in trios around slender electrical supply columns; the locations of these work stations are determined by the overhead cable suspension track that winds its way sinuously around the entire office. A large swirling drum of a conference room dominates the space, rising up to the sky like a wave in a great swell of vibrant blue and green. It is entered along the sweeping

Below: pie.com loft space

pie.com 阁楼空间

Opposite: pie.com loft space view,
the working center stage

pie.com 阁楼空间中心工作区

EXIT

Above: pie.com loft detail

pie.com 阁楼空间细部

path of a grey-on-grey floor mural, by the LA pop artist Andre Miripolsky, that leads from the elevator doors, through the entry lobby, right into the eye of the drum.

Pie.com's business focuses on dynamic, competitive, and high-intensity extreme sports, such as snowbording, surfing, skateboarding, and BMX mountain biking. Though the design of pie.com's offices includes moving parts (rolling doors and wheeled desks), it avoids being a literal manifestation of such activities. Instead, the space achieves its sense of dynamism and action through a dramatic manipulation of architectural elements. As natural light pours in from large skylights and plentiful east- and west-facing windows, it skips across exposed beams and free-standing walls, backlights translucent partitions, silhouetttes the sometimes frenetic motions of workaday goings-on, and sets various areas aglow with the reflected colors of nearby walls. This complex interplay of materials, colors, forms, and light makes for an animated environment that fosters productivity, creativity, and fruitful collaboration.

安德鲁·米罗波里斯基（Andre Miripolsky）绘制】通道，穿过入口大厅，首先映入眼帘的就是这间会议室。

Pie.com的工作主要是针对滑板、滑雪、冲浪和B MX山地自行车这些动感、有竞争性和剧烈的极限运动。虽然pie.com办公室的设计包括移动部件（滚轮移动门和带轮桌子），但是它还是要竭力避免直白地表现上述体育活动。相反，空间是通过对建筑元素的戏剧化处理而获得动感。自然的阳光透过天窗和大量东西朝向的窗户，穿越无掩蔽的横梁和自由墙，从背后照亮了半透明的隔断，勾勒出人们忙碌工作的侧影，从周围墙壁反射出的色光活跃了众多区域的气氛。这个复杂的，因材料、颜色、形式和光线之间的相互影响而创造出的生动环境，孕育着生产力、创意和富有成效的合作。

Left: pie.com loft detail

pie.com 阁楼空间细节

DESIGN
CAMP
MOBILE

Contents

海洋火车住宅

这个占地3 000平方英尺的建筑近期刚刚完成。项目使用了传统的商业、工业用材料。移动设计事务所利用在洛杉矶市中心工作现场的回收钢材和存储货柜创造了一个美丽的绿洲，他们并没有丢弃或者掩盖给设计带来灵感并提供材料的工业景观。

建筑位于一个酿酒厂旁边聚集了300位艺术家的阁楼生活工作社区，房子布满了巨大的玻璃窗，通透的空间徜徉在阳光之下，并与这个艺术家社区相通。为了与艺术家社区的精神相吻合，这个项目一开始就与客户理查德·卡尔森和建造商之间形成了一种建造与设计同步进行的试验性工作状态。

事实上，这座房屋的建造是来源于就地取材，这使该建筑与洛杉矶的工业历史完美地结合在一起。这些材料随着这个地区区域性面貌的重构而获得新生。两个酒糟槽被改造成鱼池和游泳池。几个大型存储货柜被当作创造和划分室内生活空间的主件。每个存储货柜都有自己的独特功能，一个是娱乐区兼图书室，一个是可以俯瞰下方花园的起居室和办公空间，另外两个被作为浴室和洗衣房及主卧室，一个戏剧化的重叠空间围绕着房子的上部。这个简约的空间使材料和形式得以交融，条纹金属、工业货柜和不加装饰的木材之间的视觉对比使得这个建筑融入了温暖而宁静的绿色交响。

所有被使用的存储货柜都以不同的方式得到了奇妙的改变。有些已经被分成几个部分，有些

Seatrain Residence

Completion date: June 2003
Client: Richard Carlson
Design Principal: Jennifer Siegal
Senior Designer: Kelly Bair
Assistant: Andrew Todd

Creative Director + General Contractor: Richard Carlson
Interior Design: Arkkit Forms / David Mocarski
Landscape Design: James Stone
Waterscape Design: Jim Thompson
Water Features: Liquid Works /Rik Jones
Steel Fabrication: Steel Man / Don Griggs
Glass Fabrication: Penguin Construction / Gadie Aharoni
Artist: Phillip Slagter

Recently completed, this 3,000 square foot custom residence playfully uses traditional commercial, industrial materials. Using storage containers and steel found on-site in downtown LA, Office of Mobile Design creates an oasis without abandoning or disguising the industrial landscape that inspired the design and provided the materials.

Situated by the Brewery, a 300 loft live-work artist community, the large panels of glass throughout the house open up the space, allowing natural light to pour in and connecting it to the rest of the artists' community. In keeping with the artistic spirit of the community in which this house is being built, the project has been a collaborative experiment between the client, Richard Carlson, and the fabricators using a design/build approach where creative and structural decisions were made as the house was being constructed.

This home literally grows up from the land around it, engaging with and incorporating the industrial history of downtown LA through the use of found on-site materials. Just as this area of LA has reinvented itself, so too do these materials. The grain trailers are transformed into a koi fish pond and a lap pool. The large storage containers are used to create and separate the dwelling spaces within the house. Each storage container has its own individual function, one is the entertainment and library area, another is a dining room and office space over looking the garden below, another serves as the bathroom and laundry room and yet another is the master bedroom, a visually dramatic protruding volume that wraps around the upper part of the house. This unfussy space allows for the dynamic interplay of materials and forms, the contrast of corrugated metals, industrial containers and exposed wooden beams all highlighted with warm, calm green hues.

All of the containers used in the house have been altered in surprising ways. Some have been severed into separate pieces, while others have been added onto, layered or wrapped, showing

Opposite: the front of the seatrain residence and the entry garden
海洋火车住宅前庭入口花园

Below and opposite: the runing stream in the garden was inspired by the summer Carlson spent in the Caskills, this and all other exterior water feature were created by Jim Thompson; the landscape design is by Jams Stone

住宅前庭入口花园的溪流来自卡尔森在凯斯克的夏季灵感，建筑外在的溪流由吉姆·汤姆森设计，建筑景观由詹姆斯·史通设计

the myriad design possibilities in repurposing these materials. There are wrapped design elements throughout the house including a 12-foot high steel plate fence that wraps around the entire site. At one point it lifts up, stretching to become a canopy that gives shade to the entrance, creating the feeling of the ground plane being tilted upward. Here, recycled materials are not just practical and cost effective, but they create a unique, dramatic architectural vocabulary. The innovative combination of recycled storage containers, grain trailers, steel and glass will result in a house that is highly sculptural, open and LA modern.

Left: modeling of the seatrain residence
海洋火车住宅模型

被附加、分层或者包裹。这说明了这些材料依据丰富的设计而重新定义的可能性。房屋的外围设计元素包括围绕整个住所的一条12英尺的高钢板围墙，入口处的遮篷是在围墙的一个升高点上形成的，并创造了一种地面向上倾斜的感觉。可回收材料不仅经济实用，而且还可以创造出独特的、戏剧化的建筑语言。这个把回收的存储货柜、酒糟槽、钢和玻璃结合在一起的革新方法将会创造一个高度可塑、开放和现代的洛杉矶住宅模式。

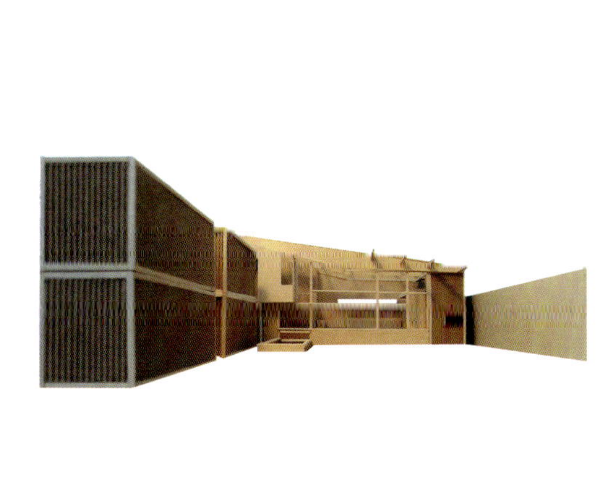

Opposite and above: interior view of the seatrain residence, the drawing is floor plan

海洋火车住宅室内和楼层平面图

floorplan 01

1 entry
2 living space
3 guest bath
4 utility + laundry
5 kitchen
6 lounge
7 bar
8 interior koi pond
9 exterior lap pool
10 library
11 media room
12 foot bridge
13 guest house

Opposite: the living area is a wide open expanse of cherry wood floors covered by rugs in muted earth tones, the furniture was all designed by David Mocarski, the expressionist painting is by artist Phillip Slagter; the drawing is house section plan

开阔的起居区地板是一色的樱桃木，上面铺有黄灰色调的地毯，所有的家具都是大卫·马凯斯基设计，墙上的表现主义绘画来自艺术家菲利普·斯莱哥特
下图是住宅剖面解析图

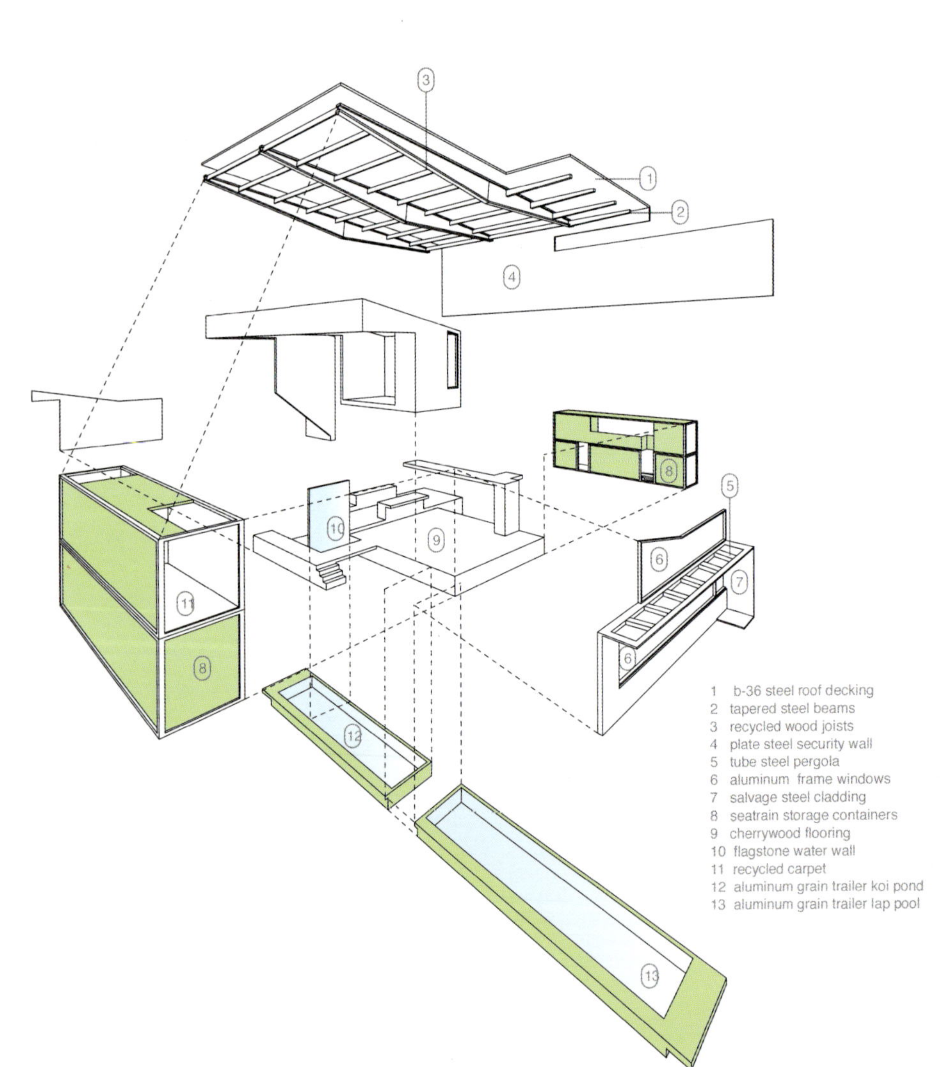

Upper: floor plan, a glass bridge on the second story of the house keeps the more private rooms seprate from the more public part of the house, a indoor lap pool is made from shipping container

楼层平面图，二楼的一座玻璃桥把私密区和公共区划分开来，室内的水池由海运货柜制成

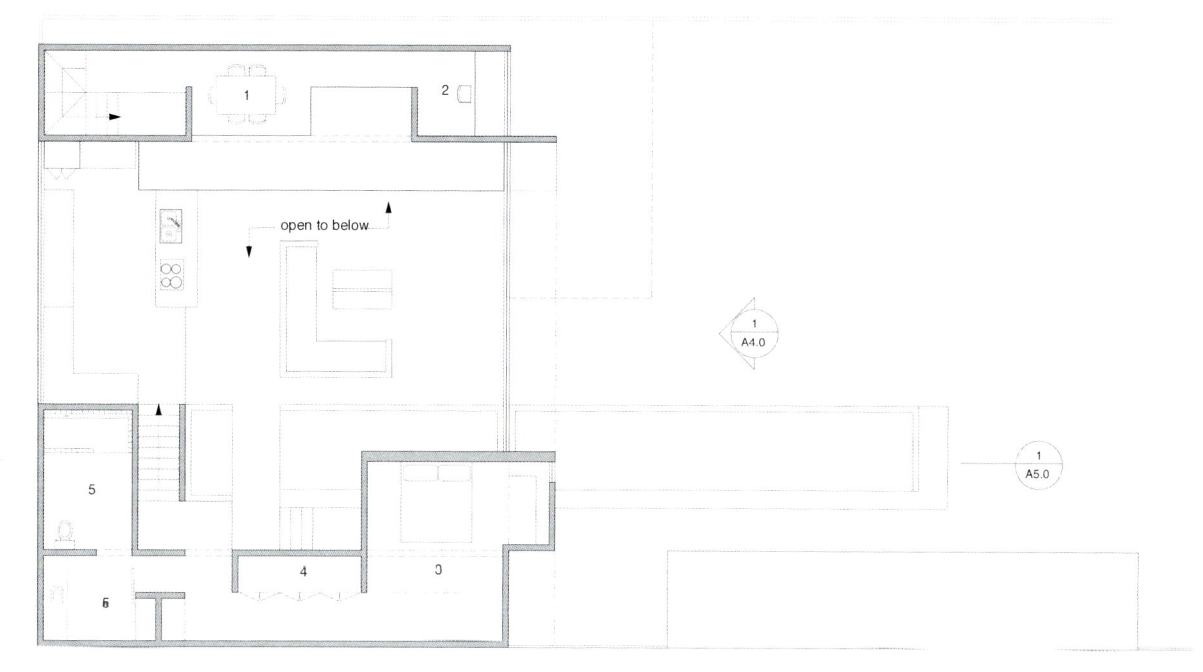

floorplan02

1 dining room
2 office
3 master bedroom
4 master closet
5 master bathroom

Right: construction sequence

海洋火车住宅施工现场

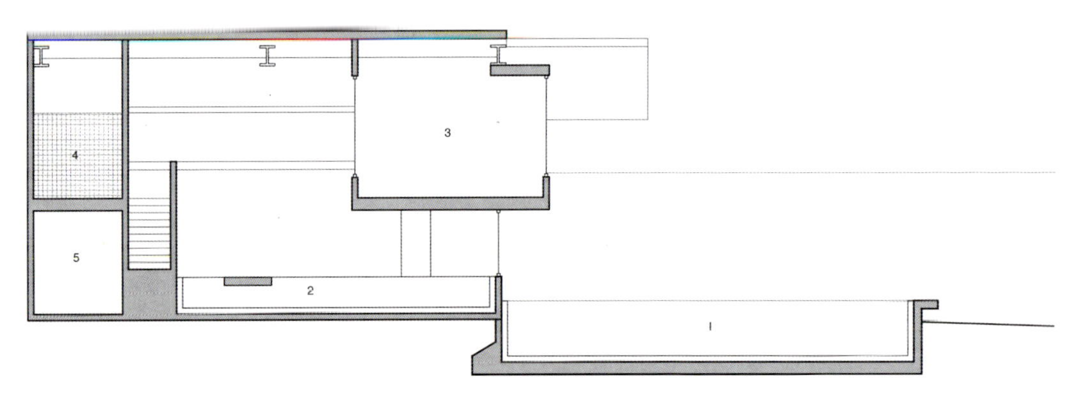

1/A5.0 section

1 lap pool
2 interior koi pond
3 master bedroom
4 master bathroom
5 library

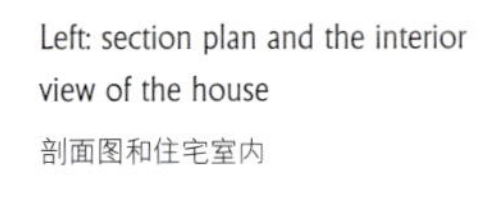

Left: section plan and the interior view of the house

剖面图和住宅室内

Opposite: view of the main living area

海洋火车住宅主起居室

Opposite: corrugated metal roofs fit with the character of the seatrain residence, as does the slate of the master plan

波状金属屋顶和蓝灰色色调是海洋火车住宅的重要特色

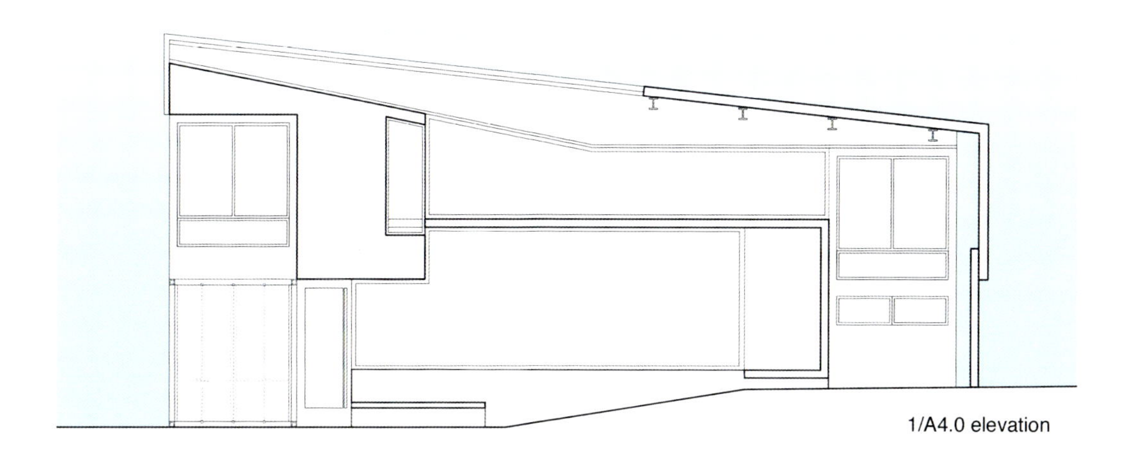

Right: elevation plan

建筑立面

Above: the bed room with a bed designed by Mocarski

主卧室中马凯斯基设计的床

Right: section plan and transverse section showing the extent to which containers, book-ending the house, have been customized

剖面图
第一层的横轴架在改造后的货柜和书架上

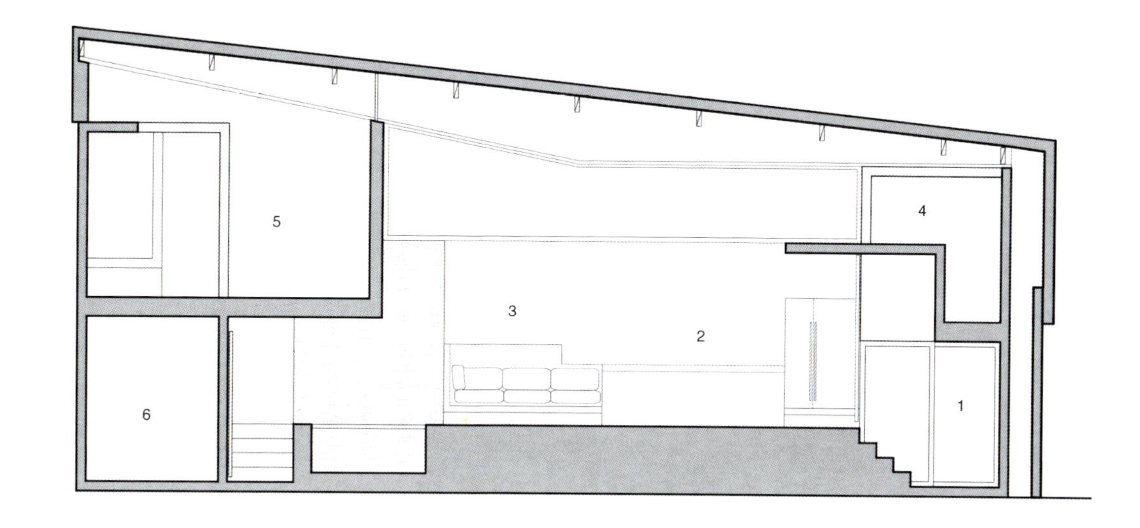

1/A5.1 section

1 utility room
2 kitchen/bar
3 lounge
4 dining
5 master bedroom
6 media room

Contents

Swell House I

Completion Date: August 2003
Clients: Dwell Magazine Invited Competition Entry
Principal: Jennifer Siegal
Design Team: Kelly Bair, Mike Dolinski, Matthew Fajkus, Simon Guest, Clay Holden, June Okado

Home ownership is the American dream, but for many it is a compromised dream when the only affordable choice is a prefab. Rethinking the prefab home is our dream and the Swellhouse brings the future to your neighborhood today.

The Swellhouse combines the smartest building technologies with cost-efficient prefabrication methods that are the vanguard of the manufactured housing industry. Championing mass-customization and celebrating individual choice, the Swellhouse distinguishes itself this field.

The signature 'S' modular structure recognizes the economy of movement where form follows necessity. Components are assembled in the factory and deployed at the site where independent 'S' modules are quickly bolted together. Electrical, plumbing, and information technologies are concealed within the cavities as part of this standardized modular system.

Likening itself to the human skeleton and smart skin, the Swellhouse employs materials for their

Opposite and below: rendering of the swellhouse

swell 住宅效果图

Right. renderings of the swell-house

swell 住宅效果图

Swell 住宅 I

拥有自己的家园是美国式的梦想。但是，当惟一的选择就是经济型的组装房屋时，这就是一个妥协的梦想了。我们的梦想是创造一种全新的组装房屋，而Swell住宅就是与你比邻而居的未来住宅模式。

在现今房屋制造业中，Swell住宅 I 号体现了将最聪明的建筑技术和低成本的预制式组装方法结合在一起的先锋技术。Swell住宅 I 号在同时适应大规模定制和个性化选择上独树一帜。

“S”形结构是对经济发展中形式适应需要的现象的认可。独立的“S”单元零部件可在工厂里生产，然后可在住宅现场快速安装。电子、管道和信息技术都被隐蔽在部件中而作为这个标准模型系统的一部分。

就像人们拥有骨骼和皮肤一样，Swell住宅 I 号也使用性能良好和环保的材料。第一层的生态阳光系统（ECOSS）玻璃板像三明治一样由玻璃、丙烯塑料和铝制天窗结合而成，能够过滤穿过幕墙的阳光并减少热量。顶层的纤维水泥板与室内墙体支撑分离形成一个防雨、防水的屏障，同时可引导自然风从墙体和金属包垫之间穿过。这种材料非常适用于加州北部的天气。

Swell住宅 I 号结合自然环境，使用了具有浸透性表面的光滑板材来建造内部庭院、外部平台或者加高的开放式入口门廊。整个规划通过寻找日常生活中的活动，比如吃饭、玩耍、睡觉、洗衣和工作之间的谐调性， 并在这些元素

performance and environmental benefits. The Ecology Sun System (ECOSS) glass panels on the first level are a sandwich of sheet glass with acrylic plastic bars and aluminum louvers, working to reduce solar heat gain by filtering sunlight from the curtain wall. The fiber cement board panels on the upper level are held away from the interior wall to produce a rain screen and water barrier while allowing natural air to flow between the wall and the cladding. This is a friendly material for the North Carolina climate.

The Swellhouse engages nature with a permeable skin of sliding panels to form interior courtyards, exterior decks or a double-height open entry vestibule. By finding harmony in the activities of everyday life - eating, playing, sleeping, washing, and working - the plan promotes a simple flow between these programmatic elements.

The Swellhouse aligns itself with the original early modernist belief that good design with environmental sensitivity through sustainable technologies will be available for the masses at low costs.

Below: house plan and exterior rendering of swellhouse

swell 住宅设计和效果图

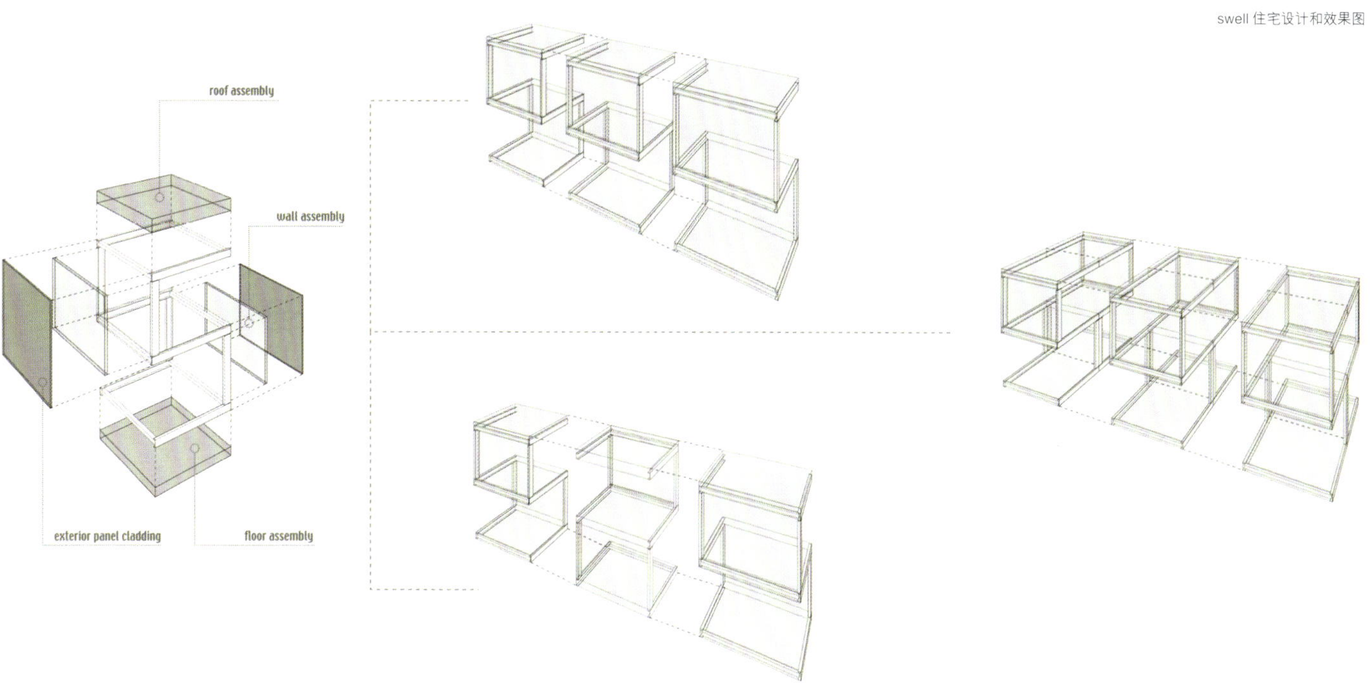

之间建立一个便捷的交流通道。

Swell住宅 I 号体现了早期现代主义的信条——具有高度环境感的好的设计通过可持续性技术将以低廉的价格为大众服务。

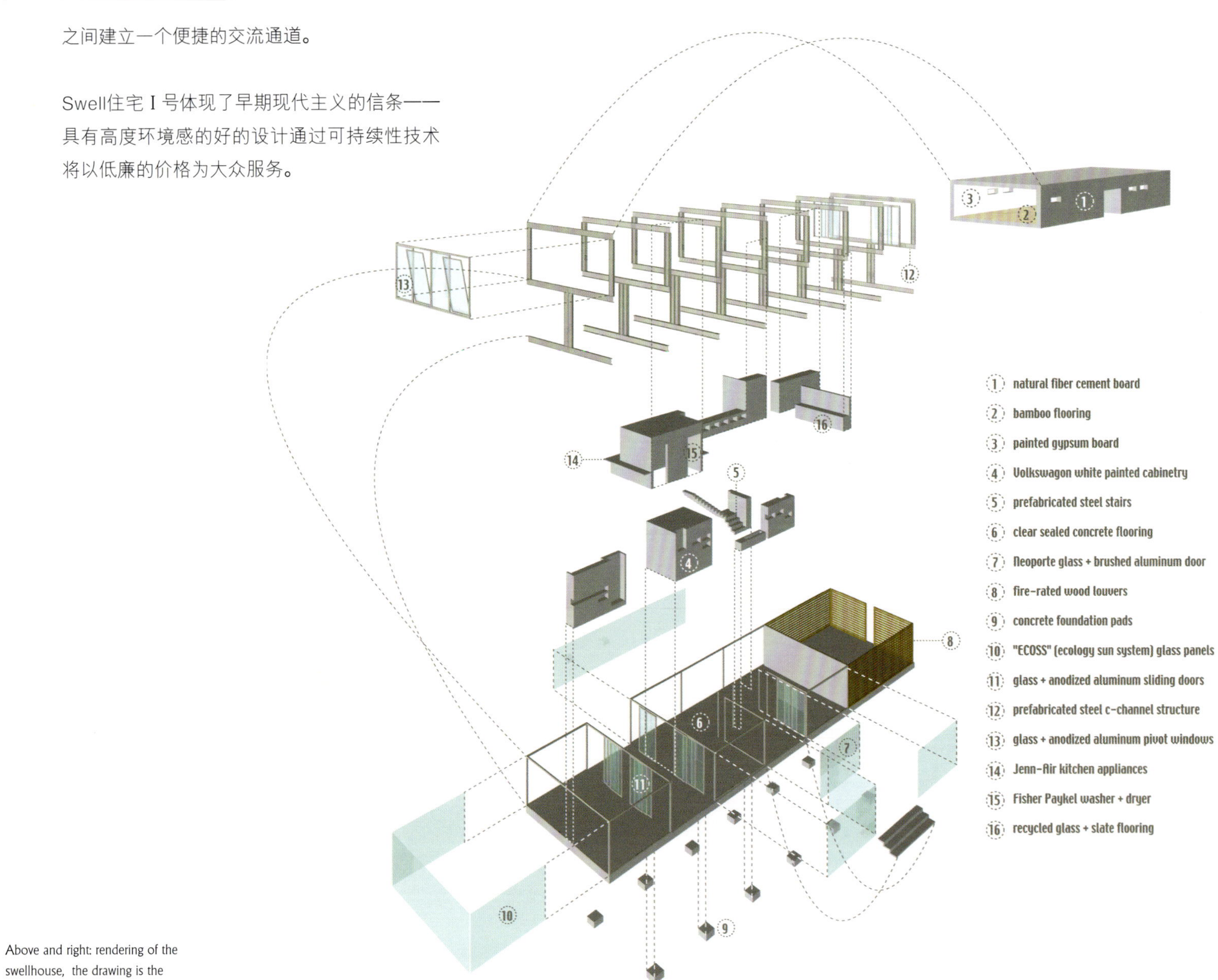

Above and right: rendering of the swellhouse, the drawing is the house plan for each section

swell 住宅效果图和布局规划图

Right: rendering of the swellhouse

swell 住宅效果图

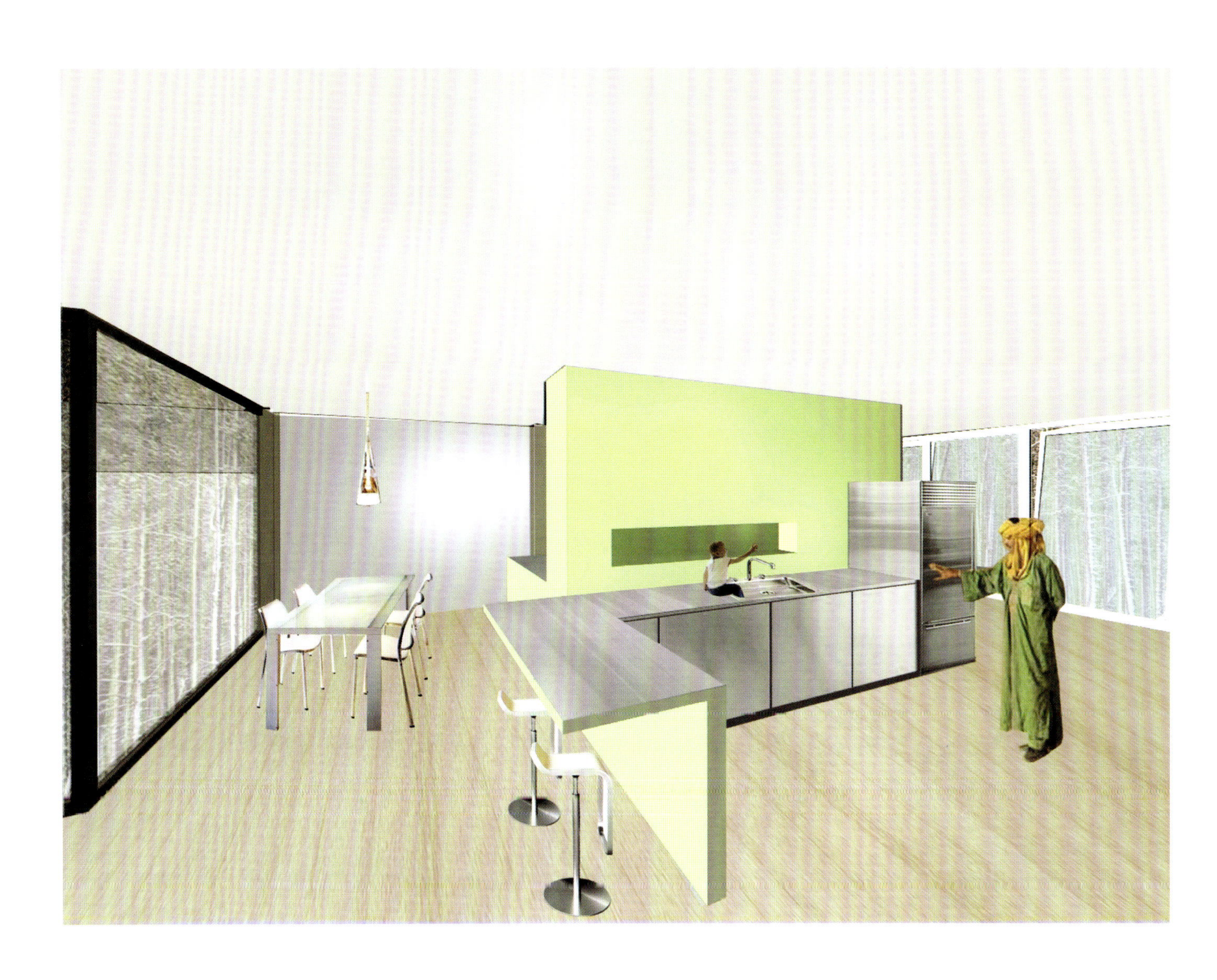

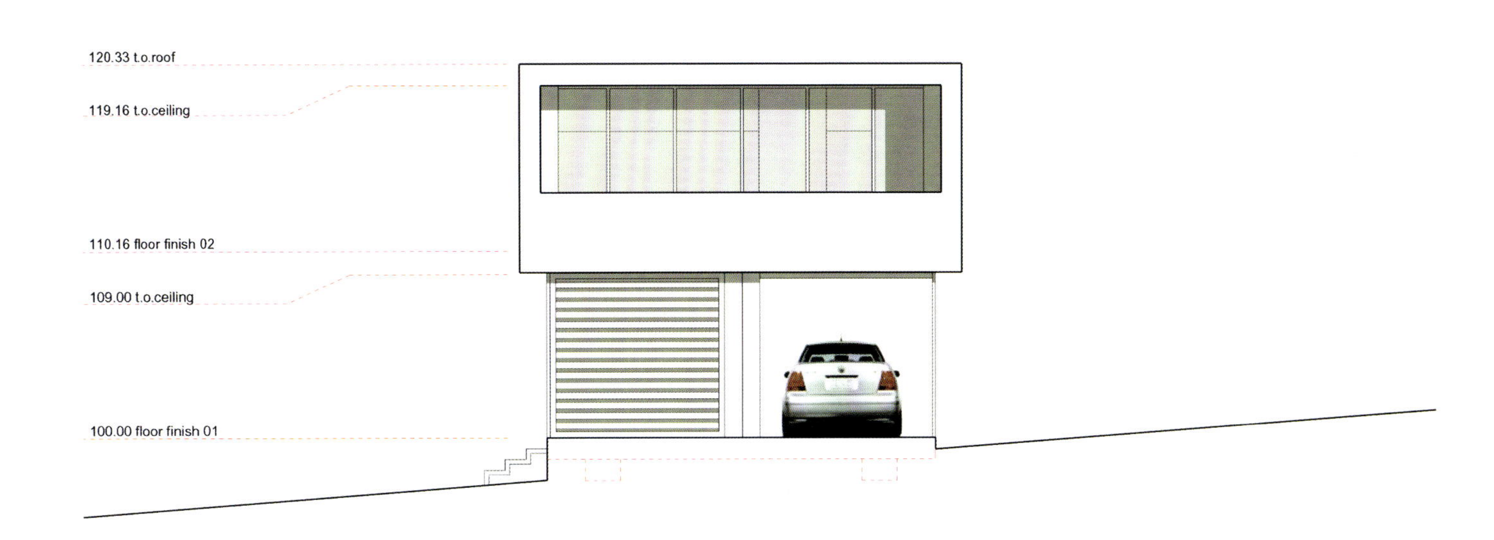

Swell House Ⅱ

Completion Date: Fall 2004
Clients: Bob and Gretchen Zwissler
Principal: Jennifer Siegal
Design Team: Nathan Colkitt,Carina bien-Willner, Tim Barnard,Barrett Cooke

Customized 3,000 square foot Swellhouse located in Manhattan Beach, California. This house is composed of the signature "S" frame with two modules intersecting creating expansive open space.

Swell住宅Ⅱ

占地3 000英尺的Swell住宅II号位于加利福尼亚州曼哈顿海滩。这座建筑由一个“S”形框架和两个交叉组件构成的延伸开放空间组成。

Opposite and right: rendering of the swellhouseII

swell 住宅 II 效果图

Contents

Tele-spot

气动空枕捕捉着网络中的图像，天空的气息在三维世界中表现着它们的奇妙，博物馆的参观者通过头顶上这个可收缩的构造物来体验时光隧道的魅力。Tele-spot是一个用于信息观察的建筑学意义上的社会容器和功能上的变量器,鼓励人们利用多重的端口进行交流，并参与到获取知识和艺术启迪的集体行动中去。Tele-spot的三个公诸于众的模型建造都使用了非常轻巧、耐用和智能的材料，极其强调动力学的应用。Tele-spot的设计源自无生命体的静态本源，它展现了变化的永恒。这是一个地平线上下两个空间的预示，每个设计不断地接纳着越来越多的使用者和一代又一代人，以及对此感兴趣者。在此，人们的需要得到了最大程度的满足。Tele-spot可以轻巧地驻足于任一选择的地点，为人们提供了一个适应性强的、可移动的资源港湾。

Right: inspirational image

tele-spot 灵感来源

Tele-spot

Completion Date: April 2001
Client: Walker Arts Center,invited competition
Principal: Jennifer Siegal
Design Team: Kelly Bair

Look up and see the sky breath. The pneumatic sky-pillows capture images from the web, representing them in the third dimension. Simultaneously, both real and virtual museum visitors experience the passage of time through the expansion and contraction of the fabric overhead. A social condenser, the Tele-spot provides a synergetic mode for viewing information. Multiple portals are provided and people are encouraged to interact and participate in the collective act of gaining knowledge and artistic enlightenment. Three prototypes of the Tele-spot are unveiled. each is efficiently constructed with extremely light, durable, and smart materials that emphasize kinetic applications. The Tele-spot design derives its origin from inanimate nature; it presents a state of permanent transformation. With an emphasis on the space above and below the horizon line, each design iteration accommodates a growing range of users, generations, and interest levels, quenching one's desire to absorb. Resting lightly on any chosen site, the Tele-spot provides an adaptable and relocatable resource port.

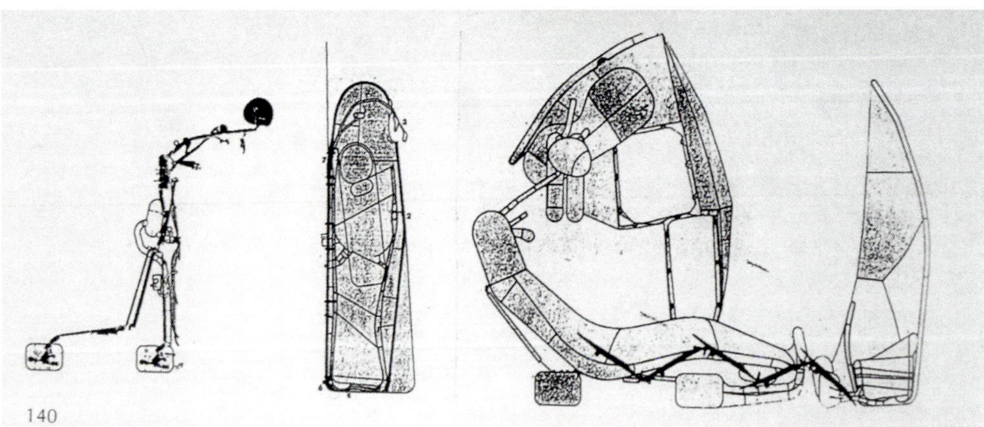

Below: rendering of the tele-spot

tele-spot 效果图

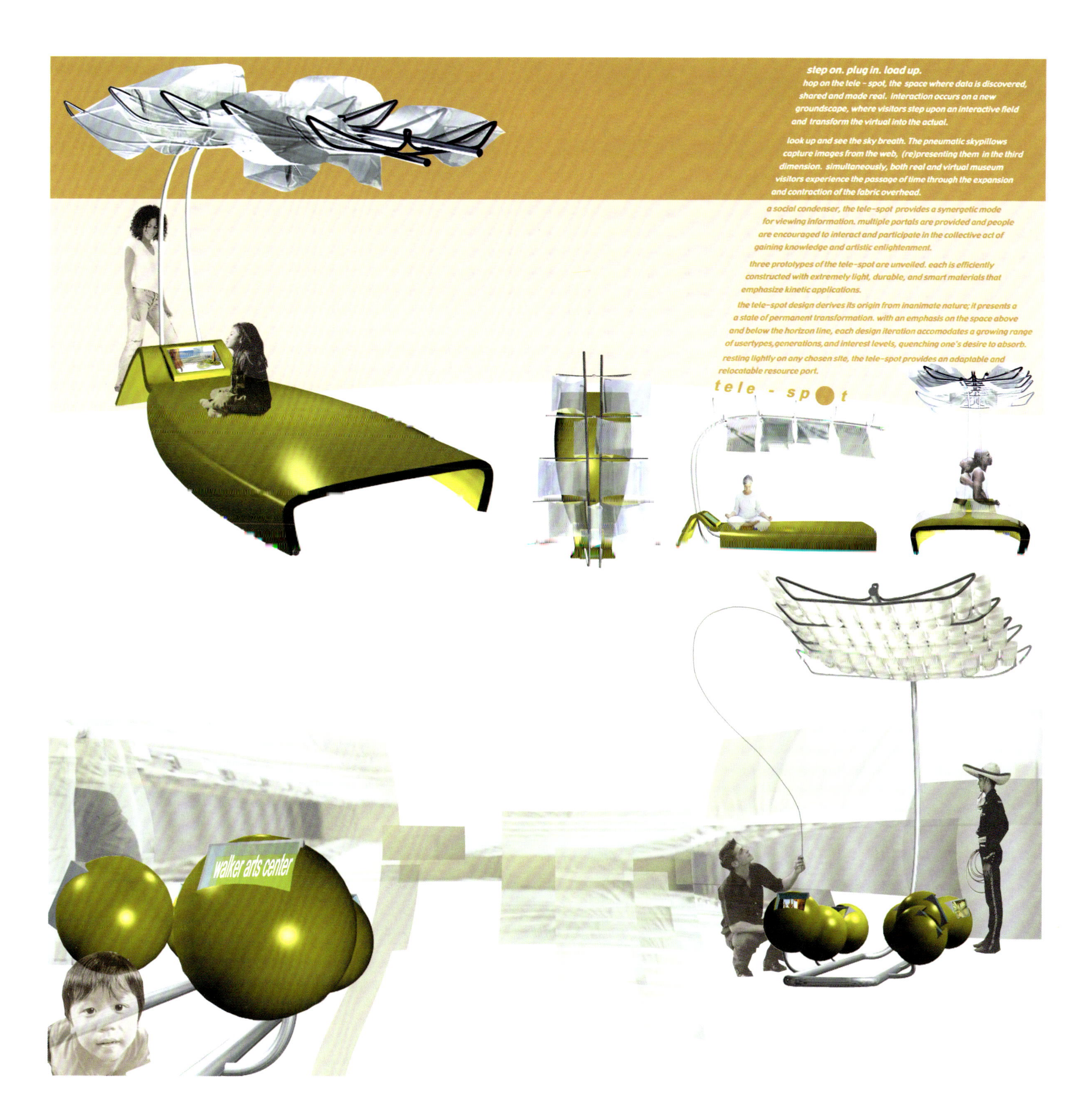

91
95
96
9
10
12-341-00

Contents

太阳大道开放实验室

“太阳大道开放实验室”位于伍德伯里大学伯班克校区建筑学院的中心位置，是一个由学生们设计并建造的占地1 050平方英尺的户外实验室。这个户外建筑为学生和教职员工们提供了一个交流、聚会的空间，还是用于学生作品展示、评判和演示的场所。太阳大道开放实验室是南加利福尼亚州最大的太阳能动力建筑之一，安装了60块由洛杉矶水电局捐赠的光电板。

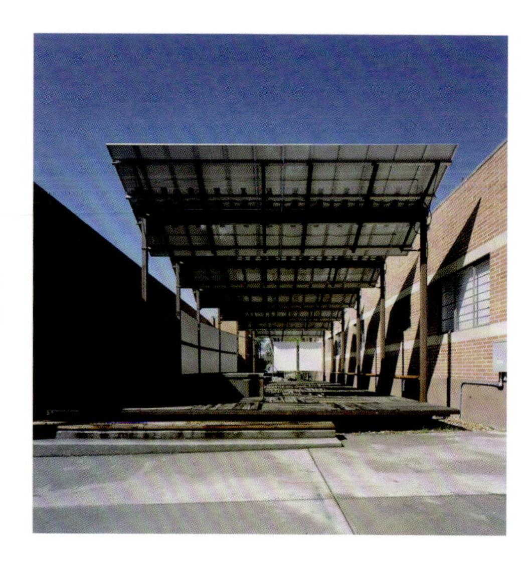

Paseo del Sol

Completion Date: December 2000
Owner: Woodbury University
Principals: Jennifer Siegal, Professor: Kevin O'Donnell, Visiting faculty

Woodbury University Student Team:
Edgar Barajas, Luc Berne, Thai Chau, Allison Fung, Pedro Jaramillo, Rubin Jimenez, Gloria Kwan, Andrie Lukman, Luis Munive, Haik Patian, Monica Plata, Somsack Rattan, David Sim Isamu Suguro

Paseo del Sol is an open-air laboratory situated at the center of the architecture school on Woodbury University's Burbank campus. Designed and built by students, the 1050 sq ft outdoor structure provides a communal gathering space for students and faculty, a garden and seating area, and a project critique space used for exhibitions, lectures and presentations of student work. Paseo del Sol is one of the largest solar powered structures in Southern California, covered by 60 photo-voltaic panels donated by the Los Angeles Department of Water and Power.

Above: exterior view of the Paseo del Sol and the student team in working progress

太阳大道开放实验室外景和学生工作小组在施工现场

Below: interior and exterior view of the Paseo del Sol

太阳大道开放实验室室内及室外景观

Contents

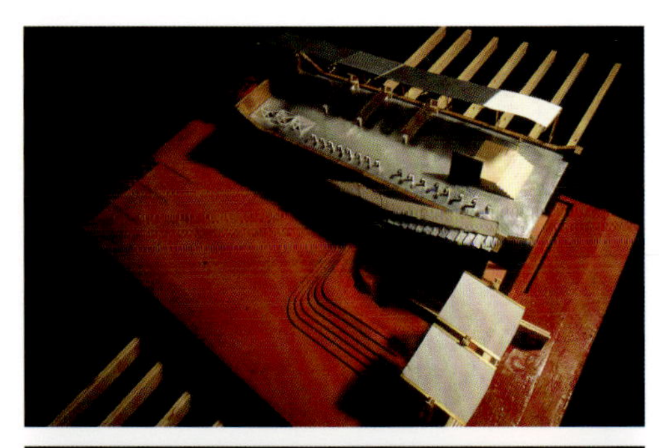

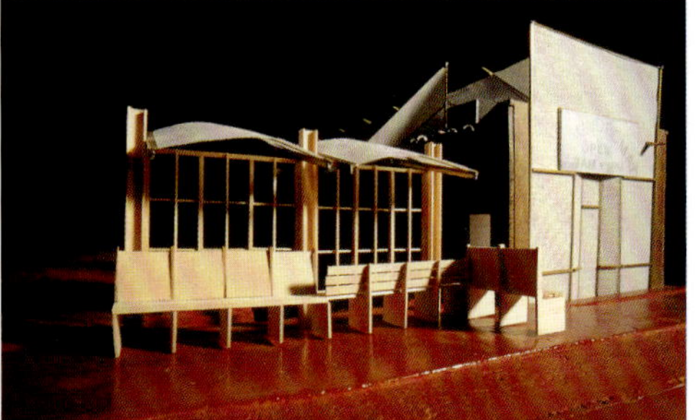

Thesis Project

Southern California Institute of Architecture

Design: Jennifer Siegal
Thesis Advisor: Hsin Ming Fung

Waiting: Times' Paradox

The thesis reconsiders the mundane architectural encounters within our daily lives. By heightening ones awareness of the "ordinary" the emphasis is shifted to the extraordinary. The idea of waiting is used as a metaphor to research physical and psychological interactions in space and how ones perception of time can be altered through spacial manipulation.

The program is both a Laundromat and a Bus Stop sited on Main Street in Venice, California. These commonly overlooked spaces epitomize the banal and make the waiting experience seem longer.

论文项目
等待：时间的矛盾性

西格尔当年的这篇论文重新考虑了我们日常生活中的普通建筑。通过提升人们对“普通”的认知，把思考的重点切入到“普通”之外的现象。等待的思想作为一种暗喻被用来研究空间中物质和精神的交流，以及人们如何通过空间的调整来感知时间的变化。

自助洗衣店和威尼斯中心大街的公共汽车站是很好的例子。这些通常被忽视的空间具有普通的特点，但能使人们获得长时间的等待体验。

Above: modelings of the thesis project
论文项目模型

Below: section plan and completed thesis project

项目剖面稿和完成建筑

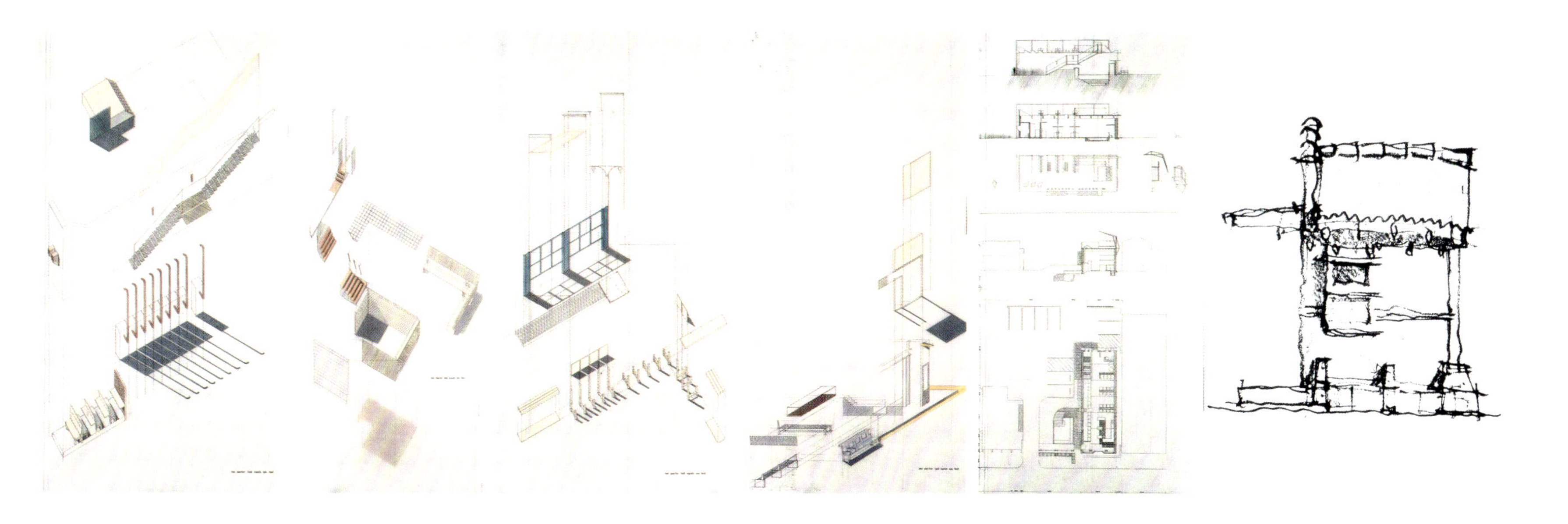

Contents

Jennifer Siegal

Jennifer Siegal is the principal and founder of Office of Mobile Design, a progressive architecture/design studio that is dedicated to the exploration and production of mobile and eco-logic structures. She earned a master's degree from the Southern California Institute of Architecture (SCI-Arc) in 1994, and was a 2003 Loeb Fellow at Harvard University's School of Design where she explored the use of intelligent, kinetic, and lightweight materials. Ms. Siegal's work, including the well-known prefab projects, Portable House and Swellhouse, was exhibited at the prestigious Cooper Hewitt, National Design Museum's 2003 National Design Triennial: Inside Design Now; and the Walker Art Center's Strangely Familiar: Design and Everyday Life. Her innovative design sensibilities and expertise in prefab and green building technologies were recognized by the popular media in 2003 when Esquire magazine named her one of the "Best and Brightest." In the same year, the Architectural League of New York included her in the acclaimed Emerging Voices program. Ms. Siegal is a Full Professor at Woodbury University in Los Angeles, and the editor of Mobile: The Art of Portable Architecture, a book and reference guide for architects on the art of transportable environments. Her forthcoming monthly publication series entitled Materials Monthly will be launched in 2005.

EDUCATION

2003
Loeb Fellowship
Harvard University, Graduate School of Design, Cambridge, MA

1994
Master of Architecture
Southern California Institute of Architecture, Los Angeles, California

1987
Bachelor of Arts in Architectural Studies
Hobart and William Smith Colleges, Geneva, New York

1986
Studies in Architecture and Art History
Syracuse University, Florence, Italy

PROFESSIONAL EXPERIENCE

1998 -Present
Office of Mobile Design, LLC, Venice, California; Founder and Principal

1995 -98
SINO Design, Los Angeles, California; Founder and Principal

1994 -95
Hodgetts + Fung Design Associates, Santa Monica, CA

1992
Mark Mack, Architect, Venice, California

1988
Skidmore, Owings and Merrill, San Francisco, California

HONORS AND AWARDS

2004
Residency, The MacDowell Colony, Peterborough, NH

2004
Finalist for Hydra21, Popular Science Magazine

2004
Faculty Design Award for Seatrain Residence, ACSA

2003
The Loeb Fellowship, Harvard University, Graduate School of Design, Cambridge

2003
America's Best and Brightest, Esquire Magazine

2003
Emerging Voices, The Architectural League, New York

2003
Honorable Mention for Portable House, Samsung Art & Design Institute, Korea

2002
Who's Who Among America's Teachers, Educational Communications, Inc.

2000

Faculty Design Award for Portable Construction Training Center, ACSA

2000

Collaborative Practice Award for Portable Construction Training Center, ACSA

1999

Westside Prize for Portable Construction Training Center, Los Angeles Westside Urban Forum

1999

Collaborative Practice Award for Mobile Eco Lab, ACSA

1999

Excellence Award for Portable Construction Training Center, ACSA/Steel Tube Institute Hollow Structural Sections Student Design & Engineering Challenge

1998

Honorable Mention for the Mobile Eco Lab, AIA Los Angeles Chapter

1997

Architect in Residence, Chinati Foundation, Marfa, Texas

1996

National Design Studio Award (jointly with Todd Erlandson), ACSA

CURRENT AND SELECTED PROJECTS

Commercial

2003

Kate Mantilini's Restaurant, Woodland Hills, CA

Clients: Marilyn and Harry Lewis

Use: 10,000 sq ft tenant improvement for trendy Los Angeles eatery.

2003

Storehouse, Commissioned for the National Design Triennial: Inside Design Now

Client: Smithsonian Cooper-Hewitt, National Design Museum

Use: A roving kiosk for displaying OMD's books, models, and images; including a 'smart' fabric skin designed by IFM.

PREFAB

THE PORTABLE HOUSE

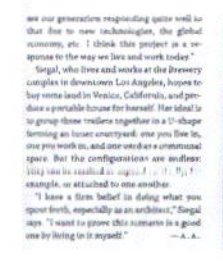

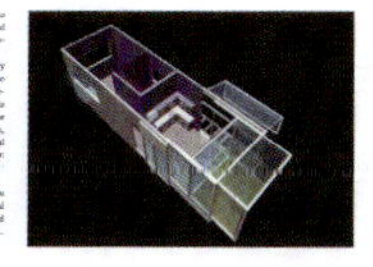

2002

Haagen-Dazs Pleasure Mobile, Invited Competition Winning Entry

Client: Haagen-Dazs

Use: A modular, portable ice cream bar and film screening environment. The Pleasure Mobile, part of a national arts initiative "The Art of Pure Pleasure," will be used at Haagen-Dazs sponsored cultural events such as film festivals, fashion shows, and musical and theatrical performances; including the Guggenheim Museum, the Seattle Film Festival, Seventh on Sixth Fashion Shows and the Chicago Art Fair.

Completed March 2002.

2001

Mobile Event City Architecture (MECA), Los Angeles, CA

Client: Pallotta TeamWorks, Dan Pallotta, Founder and CEO

Use: Overall campsite master planning for Pallotta's multiple-day events, as well as mobile structures to accommodate the events' service, transport, housing and vending needs

2000

PIE.com, Hollywood, CA

Client: PIE.com: Sebastian Copeland and Ted Owen

Use: 10,000 sq. ft. tenant improvement for extreme sports internet company.

Completed September 2000.

2000

ZEVOS Kiosk, Los Angeles, CA

Client: ZEVO Bike: Marcus Hays, CEO

Use: Electro Bike sales and service kiosk, to be franchised.

Residential

2004

Hydra 21, the ocean

Client: Popular Science magazine, Steven Madoff, editor

Use: Invited award-winning competition entry. The buoyant survival structure Hydra21 provides a temporary ocean refuge for citizens of war-torn nations.

2004

Swellhouse, Manhattan Beach, CA

Clients: Gretchen Renshaw and Bob Zwissler

Use: Invited competition entry for Dwell magazine. 3000 sq. ft. mass-customized/pre-fabricated eco-friendly house.

2004

Portable House, San Diego, CA

Client: Dr Lance Stone

Use: 720 sq. ft. pre-fabricated eco-friendly mobile house.

2003

Seatrain Residence, Los Angeles, CA

Client: Richard Carlson

Use: 3000 sq. ft. custom residence located on a 2 acre site situated at the Brewery (a 300 loft live-work artist community in downtown LA). Composed of 2 pairs of stacked ISO shipping containers (dwelling spaces) and 2 grain trailers (inside and outside Koi fish ponds) sheltered under a 50 ft. steel and glass roof membrane.

2002

ECO-Ville: Artist Live-Work Development, Los Angeles, CA

Client: Tom Ellison

Use: 2 1/2 acre ecologically sensitive urban development in downtown Los Angeles. Comprised of 40+ portable and stackable housing units deployed among flexible communal gardens and work spaces, and will include the renovation of existing warehouse structures into live-work dwellings. (under development)

2002

Hydra House

Client: Wallpaper magazine, Jonathan Bell, editor

Use: "free-form, floating structure draws inspiration both from nature and human networking."

2000

Wilshire Residence, Los Angeles, CA

Client: Michele McFaull and Dr. Peter Pekar

Use: 2500 sq. ft., two story condominium remodel in high-rise building.

1999-01 Lieber Furniture and Landscape, Beverly Hills, CA

Client: Mark Lieber

Use: Furniture Design and Fabrication, Landscape Design.

1999

Marfa Compound, Marfa, TX

Client: Jennifer Siegal

Use: Remodel of 1920's adobe structure, 1300 sq ft.

1999

Ghirardo Trailer Addition, Santa Monica, CA

Client: Diane Ghirardo and Ferruccio Trabalzi

WEEKEND JOURNAL

THE WALL STREET JOURNAL

Upwardly Mobile Homes

Can a prefabricated house be 'architecture'? Nancy Ann Jeffrey and Nancy Keates on the push to turn houses that come on trucks into objets d'art.

Use: Exterior Reading and Writing Room with Covered Deck, 300 sq ft.

1997

Chattman Compound, Cave Creek, AZ

Client: Joni and Martin Chattman

Use: Master Plan and Building Design for residential compound sited at the golden star mining company, circa 1878. The design evolution represents a collaboration between myself and Will Bruder, not built.

Educational

1999-01

Puppetmobile, California Institute of the Arts, Valencia, CA

Client: California Institute of the Arts; Cotsen Center for Puppetry and the Arts

Use: Traveling theater created for Cotsen Center director, theater students/directors and visiting artists-in-residence.

2000

Will Rogers Elementary School, Santa Monica, CA

Client: Sanford Berlin (Benefactor)

Use: Space planning and design and fabrication of furniture for multi-use computer and storage stations.

1998-00

useum of History and Perception, Marfa, Texas

Client: Chinati Foundation

Use: Temporary Museum that will serve Donald Judd's Chinati Foundation and community while support for a permanent structure is raised.

Exhibition

2004

Communities Under Construction, CityworksLosAngeles, A + D Museum, Los Angeles

2003

National Design Triennial: Inside Design Now, Cooper-Hewitt, National Design Museum, Smithsonian Institution, New York

2003

Strangely Familiar: Design and Everyday Life: Walker Art Center,

Minneapolis, MN

2002

New Nomadism: Harvard University, Graduate School of Design, Gund Hall, Cambridge, MA

2000

21 Millennium Models: selected group exhibition at Form Zero Art + Architecture Bookstore, Santa Monica, CA

2000

101 Millennium Models: exhibition at WestWeek 2000, Pacific Design Center, CA

2000

Evolution Through Practice: SCI-Arc Alumni at Work: Southern California Institute of Architecture Front Gallery, Los Angeles, CA

1999

Alumni Profile, Current Work: SCI-Arc Web Site, World Wide Web

1999

Sustainable Portables: Alternative Classrooms for Emerging Populations:

University of Southern California; Southern California Edison; Form Zero Art + Architecture Bookstore, Santa Monica, CA

1997

Radical Reconstruction: The Work of Lebbeus Woods

Architecture Gallery, The University of North Carolina at Charlotte

Curator and contractor of exhibition.

1995

The Work of Calvin C. Straub, Case Study Architect

Gallery of Design, Arizona State University

Curator and contractor of exhibition.

SELECTED PUBLICATIONS (By Jennifer Siegal)

2004

Siegal, Jennifer, Senior Editor. Materials Monthly, New York: Princeton Architectural Press, 2005 (publication date, Jan '05)

2002

Siegal, Jennifer, ed. Mobile: The Art of Portable Architecture, New York: Princeton Architectural Press, 2002

2001

Siegal, Jennifer, ed. Sustainable Portables: Alternative Classrooms For Emerging Populations, Los Angeles: Southern California Edison Press, 2001

1999 Sept.

"Mobile Eco Lab," Journal of Architectural Education, Peggy Deamer, ed., MIT Press Journals, Cambridge, MA., pp. 39-41.

1998 Spring

"Cheap (conceivably) and Temporary: In Praise of Mobility," Architecture California, Volume 19:2., W. Mike Martin, AIA, ed., AIA California Chapter, Sacramento, CA., pp. 11-21.

1997 Nov.

"Primal Adaptation: Natural Selection in Construction," Journal of Architectural Education, G. Goetz Schierle, ed., MIT Press Journals, Cambridge, MA., pp. 105-109.

1995

March "Alchemy of the Ad Hoc," Los Angeles Forum for Architecture and Urban Design, Pamphlet, Los Angeles, CA., p. 2-3.

SELECTED PUBLICATIONS (About Jennifer Siegal/OMD)

2004

Herbers, Jill, PreFab Modern, New York: Harper Design International, 2004

2004

Lacayo, Richard, "They're All Absolutely Prefabulous" Time (Summer 2004)

2004

Bissell, Therese, "Trailers For Sale or Rent" NUVO (Summer 2004)

2004

Kotler, Steven, "Moving into the Future" LA Weekly (April, 2004)

2004

"My Fellow Americans", NPR (March 17, 2004)

2004

Mani, Sujatha, "Move-Ability" Indian Architect & Builder

(February 2004)

2004

"Pre-Fab Dreaming" KCAL 9 Los Angeles News (February 12, 2004)

2004

Jeffery, Nancy Ann, "Upwardly Mobile Homes" The Wall Street Journal (Jan. 9, 2004)

2004

"Hip, Culture-Highlights, Trends and Events" Elle (Jan 2004)

2003

Microsoft, The Who's Next Issue, "Tablet PC advertisement" Newsweek (Dec 2003)

2003

Dorr, Rebecca, The Genius Issue, "America's Best and Brightest" Esquire (Dec 2003)

2003

Green, David, Homes for the 21st Century "Junk Rethunk" Dwell (Nov / Dec 2003). (Seatrain Residence featured on cover and 10 page spread).

2003

Marogna, Gege "Jennifer Siegal: Progetti Mobili" Casamica (Nov 2003)

2003

MacDonald, Heather, "Honored for 'Mobile' Ideas" LA Daily News (Nov 23, 2003)

2003

Blauvelt, Andrew, Strangely Familiar: Design and Everyday Life, Minneapolis, Minnesota: Walker Art Center, 2003

2003

Tolme, Paul, "Beyond the Trailer Park" Newsweek. (September 2003)

2003

Dr Wong Yunn Chii, Robert Kronenburg, Dr Joseph Lim eds. Transportable Environments II, London: SPON Press, 2003

2003

Kronenburg, Robert, Portable Architecture, Oxford: Elsevier/ Architectural Press, 2003

2003

Bahamon, Alejandro, PreFab, London: Hearst Book Intl, 2003

2003

Iovine, Julie V., "From High Seas to High Style" The Blade (August 24th, 2003)

2003

Iovine, Julie V., "Last Stop for Long-Haul Containers" New York Times (July 17th, 2003)

2003

Arieff, Allison, "the Dwell Home Invitational" Dwell (July / August 2003)

2003

King, Barbara, "With One Man's Vision, Nature Takes It's Course", Los Angeles Times (June 26th 2003)

2003

Vanderbilt, Tom, "The New Mobility" I.D. (May 2003).

2003

Mani, Sujatha, "Polyp abode" Indian Architect & Builder. (April 2003)

2003

Gilmartin, Ben, "Adhocism: office of mobile design" Praxis. (Issue #5, 2003)

2003

"Ecologia" Casamica. (February 2003). (Hydra House)

2003

Bissell, Therese, "The Movement Movement" HWY 111. (February 2003)

2002

Kronenburg, Robert, Houses in Motion II, London: Wiley-Academy, 2002

2002

Bell, Jonathon, "Current Affairs", Wallpaper, (December 2002). (Hydra House; "the free-form, floating structure draws inspiration both from nature and human networking.").

2002

Arieff , Allison and Bryan Burkhart, PREFAB, Utah: Gibbs Smith Publishers, 2002

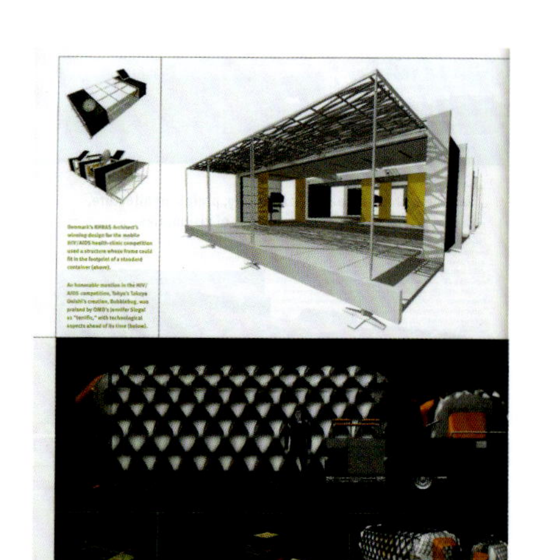

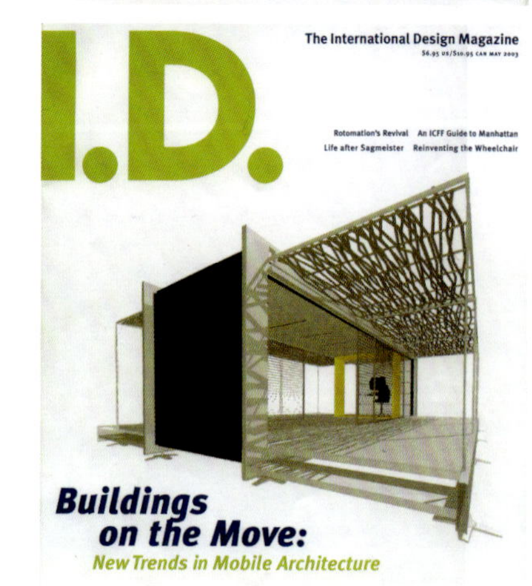

2002

Smith , Courtenay and Sean Tophan, Xtreme Houses, Munich: Prestel Verlag,2002

2002

"Digital Architect" Architectural Record. (September 2002).

2002

Fearnow, Dawson , ed. "House Proud" City. (August 2002).

2002

Bell, Jonathan, ed. "The Design Directory" Wallpaper. (July 2002).

("American hi-tech with a modest environmental edge. OMD is minimal in impact, maximal in flexibility.)

2002

McKee, Bradford, "The Impossible Dream?" Metropolitan Home. (May/June 2002).

2001

"Emerging Voices" Architectural Record, (October 2001).

(feature article on Office of Mobile Design.)

2001

Zeiger, Mimi, ed. "Yorwurts, Komraden, Wir Mussen Zuruck!" Loud Paper. (2001, Volume 4, Issue1). (feature article on Office of Mobile Design.)

2001

"Risky Business" LA Architect, (March/April 2001). (feature article on 10,000 sq ft web based extreme sports office space, PIE.com.)

2001

"Pre-Fab" Dwell, (April 2001). (3-D Computer renderings of the Portable House prototype.)

2001

Anderton, Frances, "LA Currents" New York Times, (February 2001). (3-D Computer renderings of the ZVO Kiosk prototype.)

2000

Bell, Jonathon, ed. "iMobile", Wallpaper, (December 2000).

(online news

feature of the roving cybermobile.)

BEST & BRIGHTEST CULTURE

{ Jennifer Siegal } Architecture that doesn't move—physically pick up and move—is just so second millennium. We're living in a mobile world, and if not just individuals but entire communities can't keep up when they need to, then what good are they?

2000

"Architettura Arte Povera", Lotus 105, (Summer 2000). (Photographs, drawings and article on the Portable Construction Training Center.)

2000

ARTE German/French Public Television (February 2000). (Special documentary on science and technology until 2030. Office of Mobile Design featured.)

1999

"New Nomadism" Domus. (April 1999). (Photographs, drawings and article on the Portable Construction Training Center.)

1999

"1998 Design Awards" LA Architect. (Spring 1999). (Photographs and description of the Mobile ECO LAB, Citation Award.)

1997

Carpenter, William, ed. Learning by Building. New York: Van Nostrand Rienhold, 1997. 127-130 (Illustrations and text of Primal Adaptation studio work.)

1996

Siegal, Jennifer Ruth. Mobile Laboratory: Limitations + Invention. Tempe, AZ: Arizona State University, 1996.

1995

Siegal, Jennifer Ruth. Primal Adaptation: Natural Selection in Construction. Burbank, CA: Woodbury University, 1995.

1994

Erlandson, Todd and Jennifer Ruth Siegal. 8' x 8' x 8'. Los Angeles, CA: Free Access Press, 1994.

1994

Renzis, Maria Giulia Zunina de, ed. "Studying in LA" Abitare (May 1994): 98. (Illustration of Graduate Thesis.)

1994

McConnell, Mick, ed. The Los Angeles Experiment. New York: SITES /Lumen Books, 1994. (Cited as graduate studio teaching assistant.)

1992

Lerup, Lars. men-ton med-i-a-theque, Studio Proposals. Switzerland: SCI-Arc Press, 1992. (Illustration and text

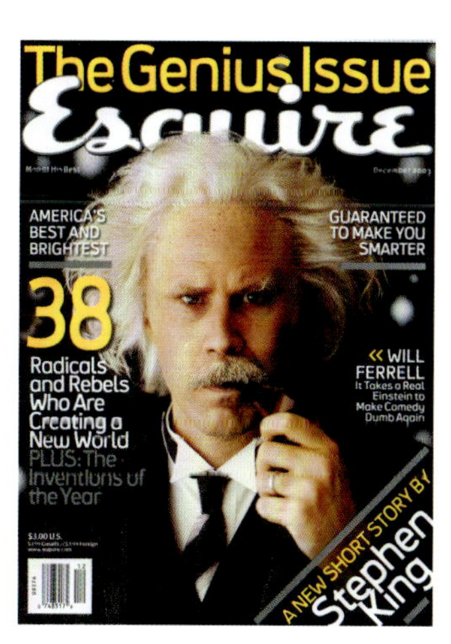

of studio design project)

EXHIBITIONS

2004

Communities Under Construction, CityworksLosAngeles, A + D Museum, Los Angeles

2003

National Design Triennial: Cooper-Hewitt, National Design Museum, Smithsonian Institution

2003

Out of the Ordinary: Design and Everyday Life: Walker Art Center (additional tour venues to be added)

2000

21 Millennium Models: selected group exhibition at Form Zero Art + Architecture Bookstore, Santa Monica, CA

2000

101 Millennium Models: exhibition at WestWeek 2000, Pacific Design Center, CA

2000

Sustainable Portables: Alternative Classrooms for Emerging Populations:

Form Zero Art + Architecture Bookstore, Santa Monica, CA

USC School of Architecture Gallery, Los Angeles, CA

2000

Evolution Through Practice: Sci-Arc Alumni at Work: Southern California Institute of Architecture Front Gallery, Los Angeles, CA

1999

Alumni Profile, Current Work: SCI-Arc Web Site, World Wide Web

1998-99 Los Angeles AIA Design Awards: Traveling Exhibition. Los Angeles, CA

1998

Mobile Eco Lab: Volume 5, Electronic Journal of Architecture, World Wide Web

1998

Learning by Building: Installation of my students' national design/build work. Store Front, Hennessey + Ingalls Bookstore, Santa Monica, CA

1997-01 Artwalk: The Brewery, Los Angeles, CA

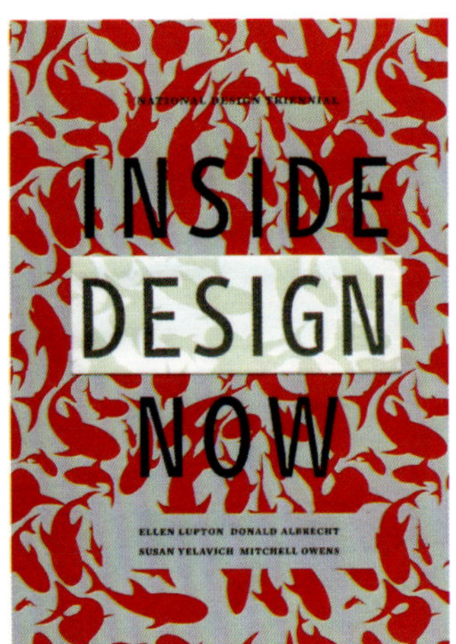

1997

threads, thresholds, and terrains: Exhibition of SCI-Arc women graduates. Southern California Institute of Architecture, Los Angeles, CA

1996

Unbuilt Architecture Design Awards: Boston Society of Architects, Boston, MA

1995

less is more: Built Work for Under $100K: (jointly with Todd Erlandson), Southern California Institute of Architecture, Los Angeles, CA

1995

Civic Innovations: Los Angeles Forum for Architecture and Urban Design, Los Angeles, CA

LECTURES AND SYMPOSIA

2003

Motopia: Washington University St. Louis, MO

2003

Where or What is Home Symposium: Panel Chair, Univ. of CA Santa Barbara

2003

Keynote Speaker, AIA Iowa Convention, Des Moine, Iowa

2003

Motopia: University of Colorado, Boulder

2003

Motopia: Massachusetts Institute of Technology, Cambridge, MA

2003

Motopia: Woodbury University, Burbank, CA

2003

OMD: University of Cape Town, South Africa

2003

OMD: The Bakery, Frankfurt, Germany

2003

Emerging Talent: Monterey Design Conference, CA

2003

Emerging Voices: The Architectural League, New York

2003

Motopia: these buildings have no borders: Ecology in Design Lecture Series, Harvard University, Graduate School of Design, Cambridge

2003

Strangely Familiar: Design and Everyday Life: Walker Art Center Minneapolis, MN (panel discussion)

2003

New Nomadism: University of Nebraska

2002

Placemakers Perspectives on Art and Architecture: Harvard University, Graduate School of Design, Cambridge

2002

Dwell Magazine's Modern Home Design in the 21st Century: 1 of 4 panelists, LA Mart, Los Angeles

2002

New Nomadism: University of California, Berkeley

2002

New Nomadism: California College of Arts and Crafts, San Francisco

Last Stop For Long-Haul Containers

2002

Meta Media, Hyper Culture: Baumer Symposium, Wexner Center for the Arts with The Ohio State University

2002

Relocation: Architecture in Transit: University of Arkansas

2001

Initiative: Architecture, Leadership and the Urban Fabric: 1 of 4 panelists, UCLA

2001

New Nomadism: University of Oregon, Eugene

2001

Office of Mobile Design: AIA San Diego/ Museum of Contemporary Art

2000

Office of Mobile Design: University of Southern California

An L.A. house showcases a new regard for industry leftovers.

2000

Sustainability and Education: Los Angeles Chapter AIA, UCLA

2000

Office of Mobile Design: California State Polytechnic University

2000

Office of Mobile Design: University of Houston

1999

The Art of Making: 1 of 6 conference presenters, University of California, Berkeley

1999

Relocation: Architecture in Transit: University of New Mexico

1999

Relocation: Architecture in Transit: Woodbury University, Burbank and San Diego

1998

Office of Mobile Design: Chatroom Host, Volume 5

1998

Portable Architecture: "Out There Doing It" Los Angeles Forum for Architecture and Urban Design

1996

Built and Theoretical Work: Powerhouse, Charlotte, NC

1996

Primal Adaptation: Natural Selection in Construction: (jointly with Todd Erlandson),Woodbury University

1996

Primal Adaptation: Natural Selection in Construction: Arizona State University

PAPERS PRESENTED AND PUBLISHED IN CONFERENCE PROCEEDINGS

2004 April

"Motopia", Design Symposium

Transportable Environments III, Ryerson University, Toronto

2004 March

"Seatrain Residence", Faculty Design Awards,ACSA 92nd

National Conference, Miami

2002 March

"Design-Build: Practice and Pedagogy", Special Focus Session

ACSA 90th National Conference, New Orleans

2001 May

"Office of Mobile Design", Design Symposium Transportable Environments II, University of Singapore, Singapore

1999 March

"Design/Build in the Curriculum"

ACSA 87th National Conference, Minneapolis

1998 Oct.

"The Mutual Attraction of Design/Build"

ACSA Western Regional Conference, Berkeley

1998 March

"Cheap (conceivably) and Temporary: In Praise of Mobility"ACSA 86th National Conference, Cleveland

1997 May

"Cheap (conceivably) and Temporary: In Praise of Mobility" International Portable Architecture, Conference and Symposium, University of Liverpool, England

1997 Jan.

"Cheap (conceivably) and Temporary: In Praise of Mobility" ACSA Western Regional Conference, Los Angeles

1996 Dec.

"Design/Build: Methodology for Education" National Building Research Institute: International Symposium "Applications of the Performance Concept in Building" Tel Aviv, Israel

1996 March

"Primal Adaptation: Natural Selection in Construction" ACSA 84th National Conference (jointly with Todd Erlandson), Boston

1996 Feb.

"Primal Adaptation: Natural Selection in Construction" ACSA Western Regional Conference (jointly with Todd Erlandson), Honolulu

GRANTS

2003

$5000 for commissioned piece; Storehouse Cooper-Hewitt National Design Museum

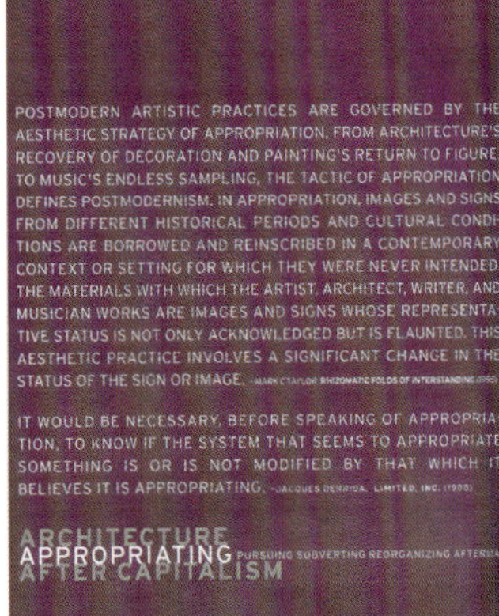

2002

$1100 Faculty Development Grant

Woodbury University

2000-01 $20,000 for Paseo del Sol, Phase 2 Facilitated and secured sponsor; Woodbury University

2000

$21,000 for Sustainable Portables: Alternative Classrooms for Emerging Populations Facilitated and secured sponsor: Southern California Edison

1999

$63,385 for Paseo del Sol, Phase 1

Facilitated and secured sponsors (major donors listed); Los Angeles Department of Water and Power ($50,000), Anvil Iron ($7500), Ferro Union ($2500), United Concrete ($1100), Carlson Industries ($500), Various Garden Nurseries ($400)

1998

$14,500 for Portable Construction Training Center

Facilitated and secured sponsors; Venice Community Housing Corporation ($2000), Westwood Salvation Army ($10,000), CBM Building Materials ($1500), Re-Sets ($500), Industrial Metal Supply ($500)

1998

$6,630 for Mobile Eco Lab

Facilitated and secured sponsors; The Hollywood Beautification Team ($1500), Carlson Industries ($2850), Re-Sets ($1500), Industrial Metal Supply ($500), IKEA ($200), Wilshire Printers ($80)

1997

$2350 Academic Program Improvement Grant

Integrating Computer Aided Design into Architecture Design Studios, The

University of North Carolina at Charlotte

1997

$3500 Junior Faculty Summer Fellowship

Funding for Chinati Foundation summer residency, The University of North Carolina at Charlotte

1996

$300 Research Grant

Herberger Center for Design Excellence, Arizona State University

1994

$1000 Publication Grant

Free Access Press, SCI-Arc

PROFESSIONAL AND CIVIC ACTIVITIES

2005

Chairperson and Program Author, Lyceum Competition (Smart Materials)

2004

Moderator, Collaborative Practice Awards, ACSA 92nd National Conference, Miami

2001-03

Board of Directors, Alumni Representative, SCI-Arc

2003

Juror, Lyceum Competition (Manufactured Housing)

2002

Juror, Architecture for Humanity (Mobile HIV/AIDS Clinic for Africa)

2002

Moderator, AIA/ACSA, Experiences in Design-Build: The Expanding Dimensions of Practice and Education, Atlanta

2001

Architectural Associate Interview Panel, City of Los Angeles

2001

Juror, New American Design Competition, Los Angeles

2000

Theme Chair and Moderator, Community and Environment, ACSA 88th National Conference, Los Angeles

1999-01

Co-President, SCI-Arc Alumni Association

Co-Curator, Exhibition Committee

1999

Plenary Session Panelist, Practice and Academia:, ACSA 87th National Conference, Minnesota

2001

Panelist, Design/Build in the Curriculum, ACSA 87th National Conference, Minnesota

1999

Design Charrette Participant, UCLA, Growing Seeds School

1999

Moderator, Engaging Community, ACSA International Conference, Rome, Italy

1999

Alumni Focus Group, Southern California Institute of Architecture

1998

Guest Curator and Moderator, Powers of Ten: Exploring the Work of Charles and Ray Eames and its Legacy in Contemporary Los Angeles, Los Angeles Forum for Architecture and Urban Design

1998

Architectural Intern Sponsor, City-As-School, LA Unified School District, LA

1997

Panelist, Craft In Architecture: AIA Georgia, Southern Polytechnic State Univ.

1996-97

Member, Women in Architecture, AIA Charlotte affiliate, North Carolina

1995

Guest Co-Curator, Architecture and Craft in Contemporary Los Angeles: (jointly with John Dutton), Los Angeles Forum for Architecture and Urban Design

ACADEMIC EXPERIENCE

1997-Present

Woodbury University, Burbank, California

Full Professor

2003

The University of Minnesota Design Institute and Target Corporation, Minneapolis, MN

Guest Workshop Instructor

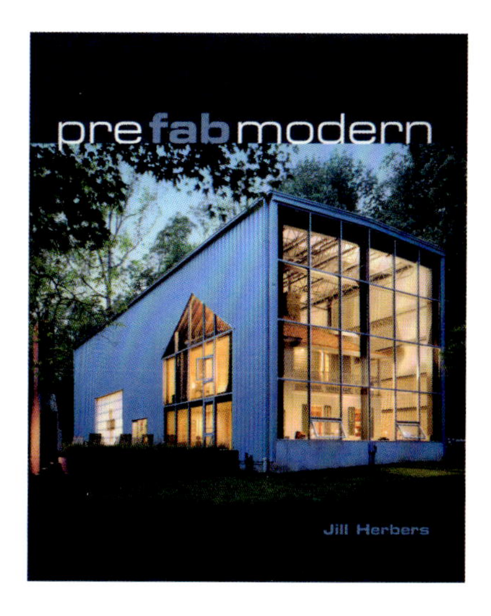

2002

New School of Architecture and Design, San Diego, California

Adjunct Associate Professor

1996-97

The University of North Carolina at Charlotte, Charlotte, North Carolina

Assistant Professor, Tenure Track

1995-96

Arizona State University, Tempe, Arizona

Visiting Assistant Professor

1995

Woodbury University, Burbank, California

Design Lecturer

1992-93

Southern California Institute of Architecture, Los Angeles, California and Vico Morcote, Switzerland

Teaching Assistant

VISITING CRITIC

2004

University of California, Berkeley, Graduate Thesis Reviews

Southern California Institute of Architecture

2003

Harvard Graduate School of Design, Cambridge

2002

MIT, Graduate Thesis Reviews

Harvard Graduate School of Design, Cambridge

Southern California Institute of Architecture

University of California, Los Angeles

Art Center College of Design

University of Oregon, Eugene

2000

Southern California Institute of Architecture, Graduate Thesis Reviews

University of Southern California, Los Angeles

University of Houston

1999

University of New Mexico

University of Southern California, Los Angeles

Art Center College of Design

Otis College of Art and Design

1998

Southern California Institute of Architecture, Undergraduate Thesis Reviews

University of California, Los Angeles

University of Southern California, Los Angeles

Art Center College of Design

Otis College of Art and Design

图书在版编目(CIP)数据
珍尼弗·西格尔／蓝青主编，美国亚洲艺术与设计协作联盟(AADCU).
北京：中国建筑工业出版社，2005
(美国当代著名建筑设计师工作室报告)
ISBN 7-112-07392-8
Ⅰ.珍…　Ⅱ.蓝…　Ⅲ.建筑设计－作品集－美国－现代　Ⅳ.TU206
中国版本图书馆CIP数据核字(2005)第042537号

责任编辑：张建　黄居正

美国当代著名建筑设计师工作室报告
珍尼弗·西格尔

美国亚洲艺术与设计协作联盟(AADCU)
蓝青　主编
*
中国建筑工业出版社 出版、发行(北京西郊百万庄)
新 华 书 店 经 销
北京华联印刷有限公司印刷
*
开本：880×1230毫米　1/12　印张：12
2005年8月第一版　　2005年8月第一次印刷
定价: **118.00** 元
ISBN 7-112-07392-8
(13346)

(邮政编码　100037)
本社网址：http://www.china-abp.com.cn
网上书店：http://www.china-building.com.cn

珍尼弗·西格尔
Report / 2005

Acknowledgements
This publication has been made possible with the help and cooperation of many individuals and insititutions. Grateful acknowledgement is made to Jennifer Siegal, for its inspiring work and for its kind support in the preparation of this book on Jennifer Siegal for the AADCU Book Series of Contemporary Architects Studio Report In The United States.

Acknowledgements by Jennifer Siegal
This book is dedicated to the next generation of global nomads.Special thanks to: Bruce Q. Lan and the United Asia Art and Design Cooperation, the fluid members of Office of Mobile Design, and my mother Gail Siegal for her encouragement, devotion, and strength.

Office of Publications:
United Asia Art & Design Cooperation
www.aadcu.org
info@aadcu.org

Project Director:
Bruce Q. Lan

Coordinator:
Robin Luo

Edited and published by:
Beijing Office,United Asia Art & Design Cooperation
bj-info@aadcu.org

China Architecture & Building Press
www.china-abp.com.cn

In Collaboration with:
OMD/Jennifer Siegal
www.designmobile.com

d-Lab & International Architecture Research

School of Architecture, Central Academy of Fine Arts

Curator/Editor in Chief:
Bruce Q. Lan

Book Design:
Design studio/AADCU

ISBN: 7-112-07392-8

出版事务处
亚洲艺术与设计协作联盟／美国
www.aadcu.org
info@aadcu.org

编辑与出版：
亚洲艺术与设计协作联盟／美国
bj-info@aadcu.org

中国建筑工业出版社／北京
www.china-abp.com.cn

协同编辑：
OMD/珍尼弗·西格尔
www.designmobile.com

国际建筑研究与设计中心／美国

中央美术学院建筑学院／北京

主编：
蓝青

协调人：
洛宾·罗，斯坦福大学

书籍设计：
设计工作室／AADCU